WHEN THE ICE IS GONE

WHEN THE ICE IS GONE

WHAT A GREENLAND ICE CORE REVEALS ABOUT EARTH'S TUMULTUOUS HISTORY AND PERILOUS FUTURE

Paul Bierman

W. W. NORTON & COMPANY
Independent Publishers Since 1923

Printed in the United States of America
First Edition

For information about special discounts for bulk purchases, please contact W. W. Norton Special Sales at specialsales@wwnorton.com or 800-233-4830

Manufacturing by Lakeside Book Company
Book design by Brooke Koven
Production manager: Louise Mattarelliano

ISBN: 978-1-324-02067-7

W. W. Norton & Company, Inc.
500 Fifth Avenue, New York, N.Y. 10110
www.wwnorton.com

W. W. Norton & Company Ltd.
15 Carlisle Street, London W1D 3BS

1 2 3 4 5 6 7 8 9 0

To my father, Joseph, who when I was growing up spent his evenings handwriting and then typing manuscripts in his third-floor office; to my mother, Babette, who first showed me rocks on the beach; to Christine, who knew ice cores long before I did; and to my children, Marika and Quincy, who have a deep love for snow and cold but, sadly, are inheriting a warming world.

Oh, Greenland, is a dreadful place,
A land that's never green,
Where there's ice and snow, and the whalefishes blow
And the daylight's seldom seen, brave boys,
The daylight's seldom seen.

—ANONYMOUS, "GREENLAND WHALE FISHERIES,"
EIGHTEENTH-CENTURY SEA CHANTEY

CONTENTS

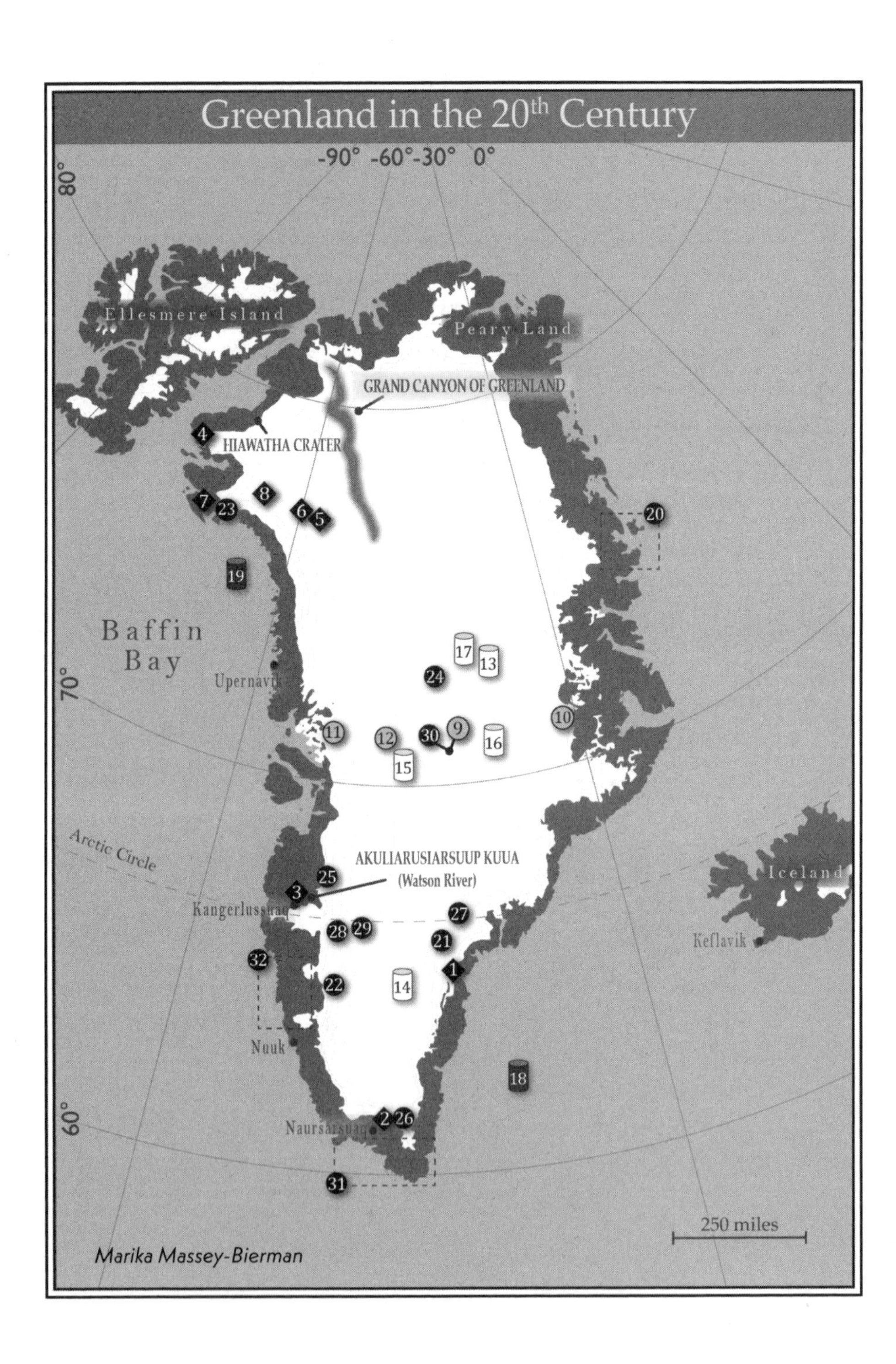
Greenland in the 20th Century
-90° -60°-30° 0°
80°
70°
60°
Ellesmere Island
Peary Land
GRAND CANYON OF GREENLAND
HIAWATHA CRATER
Baffin Bay
Upernavik
Arctic Circle
AKULIARUSIARSUUP KUUA
(Watson River)
Kangerlussuaq
Iceland
Keflavik
Nuuk
Naursarsuaq
250 miles
Marika Massey-Bierman

U.S. MILITARY BASES

1. Comanche Bay, home of the *Ice Cap Detachment*
2. Naursarsuaq / Bluie West-1
3. Kangerlussuaq / Bluie West-8
4. Site 1
5. Site 2
6. Camp Fistclench and the Underground Camp
7. Thule
8. Camp Century

WEGENER'S CAMPS

9. Eismette
10. East Coast Camp
11. West Coast Camp
12. Wegener's grave

OTHER CORING SITES

13. GRIP
14. Dye 3
15. Milcent
16. Crete
17. GISP 2 / Summit Station
18. Marine core site 918
19. Marine core site 344

LOCATIONS

20. Nazi weather stations
21. P-38s fighter plane crash
22. My Gal Sal plane crash
23. *Saviksue,* Cape York meteorite
24. Klick automated weather station
25. Site 660
26. *Brattahlíð*
27. East Greenland plane crashes in 1942
28. Project Mint Julip
29. Project Snow Man
30. *Station Centrale*
31. Norse Eastern Settlement
32. Norse Western Settlement

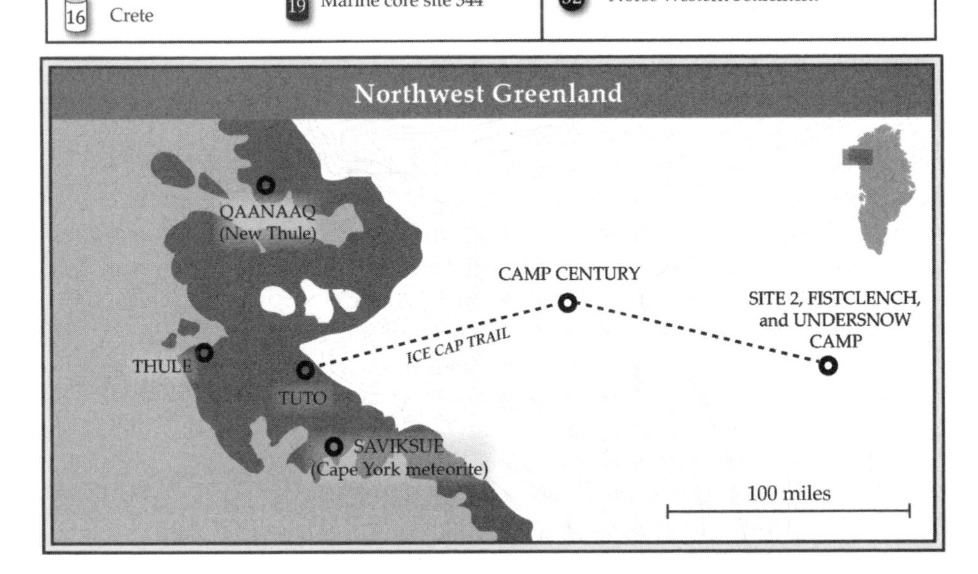

WHEN THE ICE IS GONE

Near Kulusuk in southeastern Greenland, the past winter's snow melts off the mountains as the sea ice breaks up. June 4, 2012.

PROLOGUE

> Ghost of the Future, I fear you more than any specter I have seen. . . . Will you not speak to me?
>
> —CHARLES DICKENS, *A Christmas Carol*

April 4, 2019, in Copenhagen was almost a T-shirt day. Such warmth in early spring is unusual for Denmark. But this is a country where climate change has driven up average temperatures nearly 3°F since 1875. I, though, am decidedly not warm. Standing inside a walk-in freezer bigger than many people's homes, my teeth are chattering loudly enough that I am sure the dozen or so scientists near me can hear them. It's well below zero as the fog from my breath slowly covers my beard in frost. Here, the temperature is always the same.

This glistening, windowless freezer was purpose built to store ice. Not the ice you use to cool drinks, although you could—and some have done exactly that—but long cores of ice extracted from the world's great ice sheets and glaciers. Over more than six decades, teams of drillers and scientists living and working in some of the most challenging conditions on Earth—places where bare skin freezes solid in minutes, where the sun never sets at the height of summer and doesn't rise in the dead of winter—collected the ice stored here.

I am in a rest home for these ice cores: slender, mostly transparent cylinders now cut into sections a few feet long and only inches wide. Curators here treat ice cores like rare books—with great care. They store them in meticulously catalogued boxes loaded four high on sturdy metal shelves. Laid out on the ground end to end, the ice in this freezer alone would stretch for miles—until, of course, it melted into one long, linear puddle.

Today, we are not here for the ice cores, interesting as they are. Rather, we've come to see something far less common: cores of what lay beneath the ice—brown lumps of frozen soil in vintage glass jars. If you laid these frost-mantled cylinders of earth end to end, they wouldn't even stretch the length of a car, and altogether they weigh sixty-nine pounds. My Danish colleagues have asked me to fly across the Atlantic to see these samples because I have made a long career of dating old things, especially rocks and soil from the Arctic.

To understand the coming and going of ice sheets, I analyze samples of sand in a dustless clean room and use repurposed particle accelerators from the 1960s, machines that once charmed physicists who now need bigger and better equipment. I use these instruments to weigh atoms and then count them one by one. Using such data, I've deciphered landscapes around the world, including the rocky desert of Namibia, the humid Brazilian rainforest, the dry lands of Israel, and the folded Appalachian Mountains.

Yet Greenland and its ice hold a special place in my heart. During four trips and a dozen helicopter flights there, I've seen some amazing sights, but I had never seen anything like these cookie jars holding frozen sand from beneath the ice sheet with a dusting of gravel and a hint of silt. Over my desk, there's now a photograph of that day. In it, I am holding a jar in my gloved hand, smiling.

Getting to Copenhagen, I crossed six times zones and earned four thousand, two hundred and thirty-three frequent flier miles. As one passenger on a crowded series of flights from Vermont to Denmark, I added about a thousand pounds of carbon dioxide to the atmosphere. That's unsettling, because I study when and how ice sheets, like the one covering Greenland, come and go over time as the climate warms and cools in sync with rising and falling concentrations of atmospheric carbon dioxide. Greenland's ice sheet is now shrinking rapidly, and my flight, which passed over the southern tip of the island, just helped melt a little more ice.

The jars of frozen soil in our hands came from Camp Century, a U.S. Army base hidden entirely inside Greenland's ice sheet. The camp was about 120 miles from the northwest coast of Greenland, and nearly a mile of ice separated the surface of the glacier from the rock and frozen soil below. You would not be wrong to say that Camp Century was in the middle of nowhere. In 1960, Army engineers started drilling straight down from inside the camp with the goal of collecting continuous samples of ice from the top to the bottom of the world's second largest ice sheet. After half a decade of work, on July 2, 1966, they pulled out the last sample from a depth of 4,506 feet below the snowy surface surrounding the camp.

On that day, I was four years old, the Cold War was raging, and the moon would remain free of human footprints for another three years and eighteen days. Working in frigid snow tunnels, with no heat and no daylight, the Army engineers had just finished drilling the world's first complete deep ice core. The drillers had not only sampled the entire thickness of the Greenland Ice Sheet, but had also brought up more than eleven feet of frozen soil from beneath the ice, a record that stands to this day. Nearly fifty-three years later, that material filled the jars in the Copenhagen freezer. We didn't know it then, but the soil that lay beneath the ice sheet was keeping far more interesting secrets than we ever could have imagined on that warm April morning.

FAST-FORWARD THREE MONTHS to the end of July 2019, in our laboratory at the University of Vermont. Yesterday, a van delivered two samples of the soil from below Camp Century, both still frozen solid after a several-day trip from Denmark. Two other geologists, Drew Christ and Leah Williamson, were there with me. Between us, we'd processed thousands of samples. Yet nothing had prepared us for what was about to happen.

First, we let the samples, safely sealed in plastic bags, melt slowly. Then we separated the sand from the gravel and the mud. I was hot, tired, and staring mindlessly into a bin of murky water when I saw something small and dark floating on the surface. I tossed off my safety goggles and grabbed my glasses.

"Drew" I said, "you need to see this. I think these are plant parts." From the look Drew gave me, it was clear he wasn't buying it. We were both exhausted. I'd been teaching a summer high school environmental science program for almost a week, and he was just days away from defending his doctoral thesis. I grabbed a small plastic pipette, sucked up a few of the black bits, and placed them on a white Teflon tray the size and shape of a hollow ice-hockey puck. I handed the tray to Drew. Quickly, he placed it under the microscope in the corner of the room, flipped on the glaring LED lights, and looked down the dual eyepiece. Silence. I remember best the expression of utter shock on Drew's face when he stood up and turned toward me. He glanced at Leah, same expression. Astonishment.

We took turns looking into the scope and babbling. Between the sand grains, we could see leaves and twigs—remnants of a frozen ecosystem, ancient Arctic tundra vegetation, perfectly preserved in nature's deep freeze for who knows how long. In that moment, we knew that at least once before, Greenland's ice had melted away, allowing plants to grow where today there is a mile-thick glacier. Under the microscope, we saw irrefutable evidence that ice wasn't a stable, permanent fixture on the island. Sometime in the past, Greenland's massive glacier, which today covers an area the size of France, was far smaller. I've done science for forty years and this remains my only eureka moment.

The rest of the afternoon was a blur. I brought the high schoolers in for a look, and Drew explained that these plants once sat beneath an ice sheet and that they might be a million or more years old. "Cool," they said. The Camp Century ice core was not only the first to reach the bottom of any ice sheet, but now, more than a half century later, it has a new honor—it's the first, and so far only, ice core chock full of bits and pieces of ancient plants, insects, and mosses. On that summer afternoon, two samples of brown, pebbly mud began rewriting the history of the Greenland Ice Sheet, one fossil and one sand grain at a time. The high school kids were right: this was cool, and it was only the beginning of what that frozen earth in those 1970 vintage glass jars was going to tell us. Soon, a couple of frozen lumps of soil would show the world, beset by a rapidly warming and changing climate, that Greenland's ice sheet was fragile, far more fragile than most wanted to believe. We needed to find out when it had melted, and why.

THIS IS A story about Greenland, its ice sheet, and climate change, both in the past and today. The protagonist is an ice core that radically changed our understanding of Earth, its climate, and their intertwined history. Below the ice sheet's snowy surface is another, and arguably equally important, story—one about unexpected outcomes and their consequences. No one expected climate change at the start of the Industrial Revolution. But, as the world burned greater and greater amounts of fossil fuels over the last two centuries, the impacts became increasingly clear. We know unambiguously now that warming and the loss of ice is a consequence of human actions.

When Edwin Drake struck oil in Pennsylvania in 1859, he didn't imagine the outcome would be Greenland's ice sheet melting, transferring what had been frozen water on land into the global ocean. None of the men who drilled the world's first deep ice core knew in 1966 that their work would become so important decades later, after most of them were dead. Yet a few handfuls of frozen soil from nearly a mile below those drillers' feet showed that Greenland's ice sheet was unstable even in the absence of human meddling with Earth's climate. We know now that 400,000 years ago, during an interglacial interval probably no warmer than today, tundra vegetation replaced ice without the assistance of fossil fuel–driven global warming.

Camp Century, the quirky, futuristic city that the U.S. Army carved into Greenland's ice sheet in 1960, is a critically important supporting character without which the critical ice core wouldn't exist. But, to figure out Camp Century's place in both human and climate history, as well as to understand what the science done at the camp says about our collective future, we need to go deeper—deeper into the history of Greenland and its ice sheet, deeper into the people who built the remote Arctic camp and made its ice core possible, and deeper into the Cold War that drove so much frenzied American activity in the fragile Arctic. Such context shows that the Camp Century ice core was far more than a feat of engineering: it was the beginning of a more nuanced understanding of the Arctic in general and the science of ice sheets in particular.

The chapters that follow weave together a tale of people, places, history, engineering, and science both on Greenland, a remote, largely ice-covered Arctic island, and elsewhere. That work changed the scientific world in the 1960s, and today it is changing the way we think about the unintended impacts of climate change. The story of Greenland's ice is an odyssey many millions of years long, in contrast to that of Camp Century, which lasted less than a decade.

In the Western view of things, Greenland has always been about the outer limits of human endurance, people versus nature, and meeting the challenge of the unknown and seemingly unknowable—be that by crossing the ice sheet for the first time, detecting Soviet missiles flying over the Arctic on a ballistic path to New York City, or drilling through nearly a mile of ice. A drive to know, and a certain stubbornness, kept Robert Peary and Mathew Henson, their men, and their dog teams pulling 200-pound sledges toward the North Pole in 1909, in the same way that they motivated Army drillers and scientists to spend six years coring before they finally got the equipment and techniques right and reached the bottom of the ice sheet. Human persistence in the face of challenges, both personal and professional, is central to this story.

Most people think of science as equations and equipment. But science is people, and people are unpredictable. They freeze to death. They leave their jobs. They threaten to throw an irreplaceable ice core into Lake Erie, and they misplace things, like frozen soil collected from below nearly a mile of ice. But to understand people and their day-to-day lives is to understand why the American military and scientific occupation of the Arctic happened, why thousands of men spent parts of their lives away from their families, living inside an ice sheet only 800 miles from the North Pole, and why today 8 billion people are now barreling headlong into a global climate crisis. People are the reason that the frozen soil from beneath the ice sheet vanished and then, more than twenty years later, reappeared, safe and sound in a Danish ice-core repository.

Earth is complicated. In forty-plus years of studying our planet, I've learned that the hard way. Over my career, it's become all the rage to consider Earth as a system with feedbacks that don't respond simply and smoothly to pushes and pulls. The climate system is full of jumps, times and places where temperature and precipitation patterns change abruptly. Wally Broecker, one of the first and most outspoken scientists to decipher Earth's climate system, said it well: "The Earth tends to over-respond. . . . the earth system has amplifiers and feedbacks that mushroom small impacts into large responses."[1] Greenland's ice and the frozen soil beneath it preserve the memory of such erratic behavior. They give us climate scientists pause.

Greenland's ice sheet is central to Earth's climate system and its response to human-induced climate change. When white ice and snow melt, they stop reflecting solar energy in the twenty-four-hour sunshine of the Arctic summer. Dark-colored rocks and green tundra take their place, soaking up the sun's energy and warming the planet. Water locked up for millennia in solid ice on land is suddenly freed and races to the ocean. People living by the coast 10,000 miles away from the Arctic then watch their homes go beneath the waves. Greenland has exceptionally long arms. They envelop our entire planet and its people in one way or another. Greenland's future is our future.

The more we understand about Earth, the more likely we will be to treat it better going forward. Geologists use the present to explain the past and use the past as our guide to the future. Greenland's massive ice sheet is an archive, and ice cores, like the ones from Camp Century, are frozen Rosetta stones. Once we learned how to decipher those cores, they revealed Earth's climate history, and thus a glimpse into our collective future catalyzed by seeing the past more clearly. As we face the increasing heat of a rapidly changing global climate, ice cores are key to answering one of the most pressing scientific questions of our time: What will become of Greenland's ice sheet as humans warm Earth?

Every December, for as long as I can remember, the radio station where I live has played the same recording. It's of an older Vermonter with a thick New England accent reading Charles Dickens's *A Christmas Carol.* This year, when I heard the story and noted that the grass on my front lawn was still green and bare of snow, I started to wonder if we, the people of this planet, are a collective Scrooge. If so, then the Camp Century ice core could be the Ghost of Christmas Past. But then I thought, what about the frozen plant fossils, so well preserved in the frozen soil beneath the ice? Although ancient, those bits of moss, twigs, and leaves could well be the Ghost of Christmas Yet to Come, a disturbing way for us to envision Greenland's future. A future when the ice is gone, again.

Ernst Sorge and Fritz Loewe in their bunk room cut into the snow at Eismitte on Christmas, 1930. *Fritz Philipp Loewe Collection, University of Melbourne Archives, 1988.0160.00002.*

THE INHOSPITABLE ISLAND

> Scientists still do not appear to understand sufficiently that all earth sciences must contribute evidence toward unveiling the state of our planet in earlier times, and that the truth of the matter can only be reached by combing all this evidence.
>
> —ALFRED WEGENER, *The Origin of Continents and Oceans*

Two-thirds of Greenland lies above the Arctic Circle. To the north, sea ice surrounds the land; only in the height of summer does the ice break up to reveal open water. To the south, the North Atlantic births storms that pummel the island's rocky coast with snow in the long winter and rain in the short summer. Ocean currents sweep around the island carrying armadas of icebergs. Each of these massive blocks of ice once calved from a glacier that grinds down the island's steep-walled fiords and into the sea.

Despite these harsh conditions, native people thrived on Greenland for thousands of years without the benefit of machines or fossil fuels. Three successive waves of people came to the island starting between 4,000 and 5,000 years ago, presumably from Ellesmere Island to the west in Canada—a narrow, often ice-covered crossing to far northwestern Greenland. In a place as remote as Greenland, the archaeological evidence is scarce, but scattered excavations, oral history, and more recently, DNA analyses paint an incomplete but complex picture of the island's peopling.

First came the Saqqaq, who settled the west coast of the island about 4,500 years ago. DNA of human hair recovered from a burial in Greenland's permanently frozen ground shows that these people came from the area around the Bering Sea. They were genetically distinct from Native Americans.[1] About 2,500 years ago, the Dorset replaced the Saqqaq civilization. Finally, about 1,500 years ago, the Thule Inuit, who had a genetic makeup distinct from that of the Dorset, came to Greenland from Alaska through the Canadian Arctic. Oral tradition suggests that the Thule simply overwhelmed the Dorset, who "were timid and easily put to flight."[2] One after another, the last two cultures settled around the coast of nearly the entire island.[3]

In this treeless landscape, the native Greenlanders made use of driftwood stranded on ancient beaches, now far above the waves, for fires, carving, and construction. In the winter, the Thule people, at least, appear to have lived communally in partially subterranean sod homes—a way to reduce the impact of the winter cold by relying on the thermal inertia of the ground. In warmer seasons, they traveled and hunted as small family units. Their diet varied seasonally but depended heavily on marine resources, which are plentiful around Greenland. The Thule used toggling harpoons to capture whales and made their clothing from animals native to the Arctic. They traveled quickly and widely using dogsleds and skin-covered boats called umiak.

Captain John Ross, on an 1818 search by the British for the Northwest Passage, found the northwestern Greenland Inuit equipped with iron tools, including knife blades and harpoon tips. The Inuit had cold-forged iron that they mined from the pieces of a meteorite that weighed over 145,000 pounds. The meteorite, composed of iron and nickel, is known to the Inuit as Saviksue (meaning great irons) and to meteoriticists as the Cape York meteorite. Its composition is so distinctive that tools made from Saviksue allow archaeologists to trace trade routes to southern Greenland and halfway across the Canadian Arctic. It's difficult to estimate the terrestrial age of such a large meteorite.[4] Several attempts have failed, but Saviksue probably landed when the ice sheet was larger than it is today, because there's no indication that the impact disturbed the ground beneath any of the pieces. That meant the meteorite landed at least 12,000 years ago.

Most of Saviksue is no longer in the Arctic. In 1896 and 1897, twelve years before they claimed to have reached the North Pole, American explorers Peary and Henson brought three pieces of the meteorite to New York. Henson was a Black man, a consummate explorer, and, by most accounts, reached the pole ahead of Peary. These pieces, the heaviest of which weighs 68,000 pounds, still sit in the New York Museum of Natural History. Knud Rasmussen, a Danish explorer, took a smaller piece, called Savik, to Copenhagen in 1925. Another large piece, called Agpalilik, came to Copenhagen in 1963. The last time I was in Denmark, I saw it sitting, a bit rusty and alone, in front of the Geologisk Museum. There was no sign to tell me that the rock came from Greenland or how important its metal was to the Inuit before trade with outsiders brought iron, refined elsewhere, to the island.

Norse explorers were the next people to attempt living on Greenland. The ruins of Erik the Red's farm, Brattahlíð, sit across the Tunulliarfik Fiord from the town of Narsarsuaq. When I was last there, a herd of sheep grazed around the massive stone foundation of the farm's church as a mix of rain and snow poured down. About 983 CE, Erik had arrived in Greenland with a small group of men, an advance party to investigate the island. At the occupation's peak, there were several thousand Norse living in Greenland. They settled the south and west coasts in what are known, confusingly, as the Eastern and Western Settlements. For almost four centuries, they hunted, fished, and farmed, and also traded walrus ivory with Europe.[5] There's ample written evidence of contact between the Norse and the Inuit.[6]

By 1400 CE, the Western Settlements were empty. Fifty years later, so were the Eastern Settlements. For decades, climate change took the blame for the settlements' demise. The Little Ice Age, a centuries-long cold snap during which temperatures in the North Atlantic fell perhaps by 2°F, caused the sea ice to be thicker, last longer, and spread farther south. Scientists first thought that the Norse crops failed, their animals died, and their families starved, being unwilling or unable to adapt.[7] But middens, trash heaps of domestic waste in the settlements, contained a mix of bones from land and marine animals, suggesting that the Norse adopted other food sources as farming became more difficult in the cold.

At 1300 CE, soon after Greenland's ice-core record shows that temperatures began to drop across the island, carbon isotopes in Norse skeletons confirm an abrupt change in the people's diet: the Norse had begun to eat mostly foods from the sea. As the climate cooled, the ice sheet expanded. Areas near the coast, where most Norse lived, slowly sank as the weight of the ice sheet on the land increased, and the Norse farms flooded.[8] Then, about 1425 CE, Greenland's ice-core records show that the salt content of the ice rose significantly—the fingerprint of more and probably stronger storms lofting sea spray into the atmosphere and onto the ice sheet. Navigation must have become far more hazardous. Lake sediments hint at a drying summer climate that probably also made farming more challenging. The suspicion now is that the Norse slowly abandoned Greenland and returned to Norway—their lives made more difficult, though perhaps not impossible, by climate change.

For several centuries, there was no European presence in Greenland. Then, in 1721, Hans Egede, a Norwegian clergyman trained in Copenhagen, traveled from Bergen, Norway, to the west coast of Greenland. There he searched in vain for any remaining Norse while working to convert the Inuit he encountered to Christianity. Most of his expedition died from scurvy, but Egede persevered, and founded the settlement that would go on to become the nation's capital, Nuuk. A later voyage from Denmark brought smallpox to Greenland, which killed Egede's wife and ravaged the Inuit communities.

Danish colonial rule of the island began with Egede and continued until 1953, when it became part of Denmark. The Danes granted Greenland home rule in 1979, and in 2009 again increased the island's independence. Today, a block grant from Denmark amounts to about a half-billion U.S. dollars each year, about 20 percent of Greenland's gross domestic product. Recent DNA analysis suggests that about 80 percent of Greenlanders have at least one European ancestor and that 25 percent of the native Greenlander's genome is now European—most of that Danish, and thus a testament to the interaction of the two societies.[9]

IN 1930, FIVE hundred years after the Norse left Greenland, three German men spent the winter together living inside Greenland's ice sheet. Ernst Sorge was a geologist; Fritz Loewe and Johannes Georgi were meteorologists. They were part of the last of Alfred Wegener's four scientific expeditions to Greenland—thirty men in total, the rest located in camps on the east and west coasts of Greenland. The expedition had lofty goals: to understand the ice sheet and the meteorology of Greenland.

The three scientists called their icebound camp, at 71 degrees north latitude, Eismitte (German for ice center).[10] It was 250 miles from the edge of the ice at an elevation of almost 10,000 feet, nearly at the summit of the ice sheet. From November 23 until January 20 the sun didn't rise, and in February the temperature averaged −53°F. On March 20, the mercury plunged to −84.6°F. Early in the expedition, Loewe's toes froze. One by one, his companions amputated each toe with a pen knife. Against all odds, the men survived that winter in the ice sheet. Yet, only three years later, Sorge, a Nazi Party member by then, reported Loewe, a Jew, to the police and had him jailed. After Wegener's brother got him out of jail, Loewe, fearing for his life under the Nazis, fled Germany for England and then Australia.[11]

The men used snow as a building material to create their living and working spaces. They constructed igloos on the surface and hand-dug a connected system of snow caves beneath: a living room with three bunks, a kitchen, a storeroom, a room to hold the barometer, and another to prepare weather balloons for launch. There were doors and stairs and a shaft to the snow surface for ventilation. The men learned quickly that while surface conditions on the ice sheet could be fatal in minutes, there was no wind below the snow, and the snow cave temperature, while never above freezing, was tolerable.[12] Their approach to surviving on the ice was ahead of its time. Several decades later, the U.S. Army would pick up where Wegener's expedition left off and perfect the strategy of living inside the Greenland Ice Sheet.

Wegener is best known as the father of plate tectonics, a theory based on his observation that the continents fit together like a giant jigsaw puzzle. His Greenland expeditions had an equally critical impact on science, particularly this last one. At Eismitte, Sorge, Loewe, and Georgi compiled the first climate record from the center of an ice sheet, taking weather observations three times a day for almost a year.[13] By deploying a seismograph and detonating 160 pounds of explosives, they recorded the seismic waves reflected from the rock beneath the ice sheet and used those data to measure the depth of ice beneath their camp: about 6,500 feet. Until then, no one knew the thickness of the Greenland Ice Sheet.

The most revolutionary data from Eismitte took an entire winter to collect.[14] Sorge set out to understand the history of snow deposition on the ice sheet. He hand-shoveled a sloping pit into the hardened snow below the camp; by spring, the shaft extended more than 50 feet below the glacier's surface. Sorge took more measurements each time he deepened the pit, recording the snow density, number of layers, and temperature. Between July and February, the snow temperature varied almost 60°F near the snow surface, but 10 feet into the snowpack, the seasonal change was only a few degrees. At a depth of 40 feet, the temperature was a steady −22°F. And at the bottom of his pit, the snow was nearly twice as dense as at the top—a relationship known to this day as Sorge's law of densi-

Inside the camp at Eismitte, Ernst Sorge cuts a block of snow during the winter of 1930–1931 to accurately measure its volume and weight, and thus calculate the snow's density. *The estate of Johannes Georgi, Alfred Wegener Institute.*

fication.[15] The weight of the snow above was compressing the snow beneath and, over time, slowly and methodically thinning and compacting each of the layers.

Sorge made the crucial observation that snow deposited in the winter was denser than snow deposited in the summer. He thus was able to show that the snowpack preserved a memory of the seasons and held an annual record of the past. In other words, by counting layers, Sorge could date each layer he analyzed. By the time he left Eismitte in spring 1931, Sorge had dug down through eighteen winter layers and twenty summer layers of snow. The snow at the bottom had fallen during the summer of 1910. Sorge, then thirty-one years old, was shoveling snow that fell from the sky over Greenland when he was eleven.

Sorge's approach to snow was similar to the way geologists look at sediments in a lake or the deep sea: as they build up over time, they create a layer cake of Earth's history. His was a new and transformative way of thinking about glaciers and ice sheets. Sorge's discovery meant that when people dug or drilled into the snow and ice of glaciers, they excavated history. By demonstrating that layers of snow preserve a faithful and detailed record of the past, the pit at Eismitte laid the groundwork for ice-core science, a way to decipher hundreds of thousands of years of Earth's climate history, one layer at a time.

Wegener didn't fare as well as his men. After celebrating his fiftieth birthday at Eismitte on November 1, he set off for the expedition's camp on the west coast with his twenty-two-year-old Greenlandic companion, Rasmus Villumsen. That day, the sun came up at 9:28 a.m. and rose at most nine degrees above the horizon before setting six hours later. A photograph taken before the men left Eismitte shows them wearing layers of fur clothing, standing in front of the camp with sled dogs curled up at their feet. No one saw either man alive again.

That spring, a rescue party found Wegener's body beneath a pair of crossed skis. They erected a twenty-foot metal cross to mark his icy grave, but never found Villumsen's remains. Nearly a century of snowfall has buried both of their frozen bodies, which now must lie hundreds of feet below the surface of the ice sheet. But the ice that entombs them isn't static; rather, it's slowly but steadily carrying Wegener, Villumsen, and the cross toward the lower, warmer coast. One day their remains, perhaps still clothed in the bearskins they were wearing on the day they departed Eismitte, will melt out of the ice. This is no fantasy: in 2017, a couple entombed in ice for seventy-five years melted out of a glacier in the Swiss Alps. They still had hobnailed boots on their feet.[16]

THE UNITED STATES had little interest in Greenland before the Nazis invaded Europe. With the passage of the Lend-Lease Act in March

1941, the Americans started supplying the Allies, mostly Britain and the Soviet Union, with combat aircraft. With U-boats prowling the Atlantic, ocean shipping was too risky. The military chose to fly the planes to Europe, which meant refueling in Greenland because most aircraft did not have sufficient range for a direct flight over the ocean. In April 1941, Denmark signed a treaty allowing the United States to build bases on the island. By June, construction began.

The "Snowball Route" was a three-hop air path from North America to Britain.[17] It was 780 miles from Goose Bay, Labrador, northeast to the first refueling stop in southern Greenland, Narsarsuaq, referred to by the American military as Bluie West 1. The second stop, the Keflavik airfield in Iceland, was another 750 miles east. From there, it was 850 miles southeast to Prestwick in Scotland. In summer 1942, Paul Tibbits safely delivered a B-17 Flying Fortress bomber to Europe.[18] Three years later, he piloted the *Enola Gay* and dropped the first atomic bomb on Hiroshima. Hundreds of planes attempted the journey up Greenland's fiords and over the ice in 1942. Military aircraft of the time could manage these distances, but barely, with little fuel in reserve for when things went wrong, as they often did.

Without radar, good maps, and other navigational aids, many airmen and their planes didn't make it across the ice sheet. In July 1942, a half-dozen P-38 Lightning fighters encountered bad weather flying to Iceland and, running out of fuel, set down on the Greenland Ice Sheet, accompanied by two B-17s. Rescuers retrieved the men, but the Army abandoned the planes. That same summer, three more B-17s went down when bad weather prevented them from landing at Narsarsuaq.[19] One, *My Gal Sal*, landed on the ice sheet about fifteen miles inland from the ice margin and a hundred miles south of the other large U.S. airbase at the time, Bluie West 8, which is today Kangerlussuaq Airport.[20] In October 1942, more than 5 percent of the warplanes flying to Europe crashed.

The paucity of weather data was particularly devastating to pilots and planes. Without weather stations on and around the ice sheet, aircrews and rescue teams were flying blind into fog, low cloud ceilings, and storms. Knowing the weather in Greenland was also critical for making accurate forecasts in Europe, as most weather came to the war zone from the west. Nazi Germany, east of the Atlantic, had much less knowledge of the coming weather than the Americans. The solution for both sides was building a series of covert weather stations—an effort referred to by some as Greenland's weather war.[21]

In 1942, the Nazis built four manned stations on East Greenland. One of them intentionally sent a coded message to American pilots directing them into storms.[22] The Nazi weathermen worked undiscovered in Greenland until 1943, when the multinational East Greenland dogsled patrol found their stations. Soon after, the U.S. Coast Guard captured sixty Nazis, most of whom were weathermen, not soldiers. Nevertheless, the incident made for excellent newsreel footage.[23]

The Americans built a network of weather stations around the southern Greenland coast and relied on information from Danish stations from the rest of the island.[24] But there were no data from the ice sheet, a major weather maker. In 1942, twenty-nine American men, known as the Ice Cap Detachment, went to southeastern Greenland to fill the forecasting gap.[25] They set up a base camp by the coast at Comanche Bay, along the air route between Narsarsuaq and Keflavik. They had orders to establish three weather stations high on the ice sheet—each with a four-man crew housed in a permanent building designed for burial by rapidly accumulating snow. When not recording and reporting the weather, they scouted landing strips for airplanes on the ice and rescued the planes that continued to go down on Greenland's ice sheet. Alistair MacLean's adventure novel *Night Without End* seems loosely based on the Ice Cap Detachment although it's set fifteen years later.[26]

The U.S. Army provided the Ice Cap Detachment with four-person, gas-powered tracked vehicles that could move across snow and ice. These vehicles, known as Weasels, were built by Studebaker, and were steered with two control levers, one for each track. Engineers crafted the prototype in sixty days. This fast-tracking had consequences: the Weasels often broke down, leaving their occupants stranded. In East Greenland, airdrops of replacement transmissions and differentials, allowing repairs on the ice, were common. Despite this mechanical frailty, the men traveled 150 miles by Weasel to the ice sheet's summit, elevation 10,000 feet—a revolution in over-snow travel.

For the Ice Cap Detachment, November 1942 was a brutal month. First, an Army DC-3 went down on the ice sheet near the East Greenland coast. The five crew were alive and the plane intact. The crew radioed their position and sent up flares. A rescue team set out on crude motorized snow sleds that soon broke down. A B-17 bomber searching for the lost plane encountered a whiteout, where snow and sky were indistinguishable. It crashed when its wing tip hit the ice sheet. Repeated airdrops of supplies kept the B-17 airmen, stuck on the ice sheet, alive.

Then, an amphibious one-engine biplane, a Grumman Duck, did something no other plane had ever done. It belly-landed on the ice sheet and took off again, bringing out two injured crew from the B-17. On a second rescue mission, the weather closed in, and the Duck crashed. Rescuers spotted the tangled wreckage, but there was no sign of life.[27] Over the next weeks, there were more attempted rescues, more planes went down, and a motorized sled tumbled into a crevasse, killing the driver.[28] On November 22, 1942, the Army closed the Snowball Route for the winter.[29] Rescuers never found the five-man crew of the DC-3, but more than a dozen other men suffered through a dark, ferocious Greenland winter living in snow caves or in the wreckage of their planes.

The conditions were brutal. As one of the men recalled, "Through a hole in the top of the tent they stuck an ice axe so they could get air. They crawled into their sleeping bags. . . . On Christmas eve, they had dehydrated beans and melted snow for dinner. The next morning, they said Merry Christmas to each other. . . . Outside of that nothing much happened. They just hibernated."[30] Neither the downed airmen nor their rescuers had the skills or equipment to maneuver successfully in the Arctic winter or reliably survive on the ice sheet. Ice and frozen ground were nothing like the theatres of operation they understood. Writing in a magazine for soldiers, H. H. Oliphant suggested that, "The equipment supplied to the unit was woefully inadequate. Misfortune persistently dogged the heroic efforts of this little band."[31]

The Ice Cap Detachment had demonstrated just how little the U.S. military knew about operating in the cold, icy, and unforgiving environment on and around the Greenland Ice Sheet. To successfully live, maneuver, and fight in the Arctic, the U.S. military had to understand ice. In pursuit of that goal, the Army kick-started the science and engineering of frozen places.[32] Success meant predicting Arctic weather, the characteristics of snow, the flow of ice, and the behavior of glaciers.

In the two decades after World War II, the U.S. armed forces built both the practical and intellectual foundations needed to understand, and thus work effectively in, cold regions. Out of that expertise, and out of the vision of both scientists and engineers supported by the Army, came the techniques of ice coring and the science of interpreting ice cores. The collection of the world's first deep ice core at Camp Century, Greenland, was the culmination of a far broader Cold War effort to understand, operate in, and defend cold places.

TODAY, GREENLAND IS still not an easy place to live. Always, there's the cold. The record low temperature occurred on December 21, 1991, at a camp called Klinck. It's an automated weather station on the top of the ice sheet, at about 10,000 feet elevation. It was −93.2°F—still, that is a balmy 35°F warmer than the world-record cold recorded in Antarctica. Most of the life on Greenland, including people, dwells away from the margins of the ice sheet and close to the ocean, where conditions are more moderate.

While trees are rare on Greenland, a few do grow in southern valleys, sheltered from the wind. Others have been planted in towns and arboretums. Elsewhere, along the edges of the island, outside the margins of the ice sheet, lies Arctic tundra. Here, stunted woody shrubs eke out an existence, rising to heights somewhere between your ankle and knee. Their roots tightly grip the stony soil, much of which remains frozen for all but a few months of the year. The wind beats the miniature trees down and contorts their trunks. After years of such abuse, the branches mostly lean away from the wind.

The ice sheet has shrunk as the Little Ice Age, which catalyzed the Norse abandonment of Greenland, waned and the Industrial Revolution spewed climate-warming greenhouse gases, including carbon dioxide and methane, into the atmosphere. As the ice retreats, soil slowly develops, and with it come more plants. As decades have passed and some areas have gone without ice, life has slowly established a foothold on the once barren landscape—some moss, a bit of lichen, a blade of grass, and insects.

In the spring and early summer, as snow melts and water runs over the bare rock and saturated soil, those insects hatch and swarm through the air. Even the smallest patch of bare skin is fair game, as I unfortunately learned firsthand on a hike during an unusually warm July in the field. When I came across a lake just a few feet from the trail, it seemed like the perfect opportunity to cross swimming in Greenland off my bucket list. I peeled away my hiking clothes, thinking the cold water would be my worst enemy. It wasn't. The cloud of bloodthirsty bugs surrounding me as I climbed out of the water was.

Waste disposal is a modern problem in Greenland and a greater concern. In many towns, sewers are not an option. Much of the landscape is rock, and what little soil exists is largely permafrost—permanently frozen ground. I spent a week of my first field season in Greenland based in the island town of Upernavik, living in a well-equipped, modern home. The urinal next to the toilet emptied through a pipe onto the lawn, and the toilet held a bright yellow, very heavy plastic bag. Every few days, we set the yellow bag by the roadside, where a slow-moving loader picked it up and took it to the dump. The toilets themselves were empty shells of porcelain that never flushed.

More than ninety years after Eismitte, communication to and from Greenland can be surprisingly good. There's cell coverage in all the major towns. Most field parties in the Arctic carry a satellite phone. While doing fieldwork in Greenland, we called the U.S. National Science Foundation headquarters at Kangerlussuaq every morning to tell them we'd made it through another night intact. Once a week, we had ten personal minutes of satellite phone time. I called home from remote towns of 200 people, and I called from the ice sheet. It was hardly more difficult than dialing a cell phone number with a few extra digits. But the lag time, sometimes nearly a second because the satellites orbit so far above Earth, had people talking over each other.

The daylight, or lack of it, can be much harder to handle. Above the Arctic Circle, there are days when the sun never rises and days when the sun never sets. The farther north one lives, the more extreme the daylight change is. The long polar nights and the twenty-four-hour summer days played havoc with men from farther south living temporarily in the high Arctic. In 1954, Eldon Severson, an American pilot stationed in far northern Greenland, wrote, "It's been said that Thule [Greenland] is a two day tour, one day and one night. From November to February aircrews log night [flying] time only. The sun just doesn't bother to come up for three months. From late April to August, the reverse is true. It is strictly a daytime operation. You can pick up a pretty fair suntan at three o'clock on a July morning."[33]

The wind in Greenland can be deadly. It comes from North Atlantic storms and their interaction with the ice sheet. Air speeds up as it moves over the ice and the coastal mountains. Such high winds can blow the snow into multistory drifts. On March 8, 1972, the worst storm on record struck northern Greenland. At 207 mph before the wind gauge blew away, it holds the world record for the highest wind speed recorded near sea level. The local news station reported that "baseball size rocks were picked up off the ground and flung by the wind into buildings."[34]

GREENLAND IS IN many ways a field geologist's paradise. Glaciers have scoured the landscape of soil, leaving mostly bare rock exposed at the surface, which is easy to map and sample. Solid outcroppings of rock smoothed by the passage of glaciers are as common as hillsides studded with boulders or valley bottoms filled with sand and gravel. There are no impenetrable forests and few cities with paved-over outcrops. On the other hand, ice covers 81 percent of the island, leaving geologists to connect the dots from one side of the ice sheet to the other with little information. And without

A snowstorm at Thule Air Base left this snow-choked doorway in its wake. First Lieutenant Larry Biddison of Palo Alto, California, uses the snow as an auxiliary beverage cooler. Photograph published January 6, 1961. *Milwaukee Journal, © Fred Tonne—USA TODAY NETWORK.*

a boat or a helicopter, or both, it's tough to do fieldwork. But the rocks, where you can see them, are fascinating.

Plate tectonics is responsible for the rocks of Greenland. Its folded and faulted outcrops, so clearly visible from the air and on the ground, preserve the roots of long-eroded mountain ranges that came and went over several billion years. Rocks composed of sediment, such as sandstone and quartzite, preserve material shed from those now-vanished highlands. Dark basalt covers some of the landscape, left over from volcanoes that erupted as the North Atlantic Ocean formed and Greenland separated from Europe and North America between 55 and 65 million years ago. Greenland's rocks hold mineral riches: gold and diamonds, as well as the world's largest deposit of cryolite, which for much of the twentieth century was an irreplaceable ingredient in the production of aluminum—the strong, lightweight, strategic metal critical for making aircraft.

Nuuk, the largest city in Greenland (population 19,282 in 2023), sits on the world's oldest rocks, formed between 3.8 and 3.7 billion years ago—a mere 700 million years after Earth itself formed. Among them are volcanic rocks, including pillow lavas formed by underwater eruptions, and sedimentary rocks, indicating that erosion was already shaping the planet's surface. Metamorphism heated and squeezed these rocks as Earth's continents crashed together. Geologists refer to this collection of rocks as the Isua Supracrustal Belt.

Knowing the Isua rocks were old, many scientists hoped they could find hints of ancient life in the preserved sediments, and so spent years probing them. They came up empty handed. Metamorphism had twisted any fossils in the Isua rocks beyond recognition and had changed the rocks' chemistry. Still, people continued looking, hoping they'd somehow find biosignatures—signs of ancient life.[35] The disappearance of a long-lived patch of snow, perhaps because of the warming climate, changed the odds. That melting revealed a new outcrop of Isua rocks made of dolomite, a sedimentary rock rich in magnesium, calcium, carbon, and oxygen. Unlike the rocks around it, the dolomite had evaded metamorphism. The outcrop was only a few hundred feet long, but it had uniquely preserved the original layers of sand, indicating that the rock remained in its original, unaltered state.

Within the dolomite, geologists saw stromatolites, distinctive layered structures created by communities of aquatic microbes. This discovery meant that the Isua rocks hosted the oldest direct evidence of life on Earth. In 1996, an international team of scientists analyzed minuscule amounts of carbon hidden inside some of the individual mineral grains from these rocks. They found a relationship between the two stable isotopes of carbon that only biotic processes—the signature of life—could have created. Now, they had firm isotopic evidence for the emergence of life on Earth by at least 3.8 billion years ago.[36] Today we see stromatolites forming in warm, shallow, salty lagoons, such as those along the coast of western Australia. Such environments don't exist in the Arctic, suggesting that the rocks in Greenland originated at much lower latitudes. Plate tectonics not only metamorphosed the Isua rocks, but also moved them thousands of miles poleward from where they had formed.

GREENLAND'S ROCKS GIVE a glimpse into the island's long history, but cores of ice and mud collected from the deep sea better tell the story of its ice sheet. In late October 1993, the internationally funded drill ship *JOIDES Resolution* was in the open ocean southeast of Greenland, seventy miles from the coast, where the water was just over a mile deep. The nearly 500-foot-long ship was purpose built to retrieve long cores of rock and sediment from the ocean floor. By early November, the ship's crew had drilled and recovered over 4,000 feet of core, including sediments that preserved a history of Greenland stretching back 40 million years.

Seven months later, a paper titled "Seven Million Years of Glaciation in Greenland" changed how scientists thought about Greenland and its ice.[37] While analyzing the deep-sea core he'd helped to collect, Hans Christian Larsen, a Danish geologist, found the distinctive signature of glaciation: dropstones. As ocean currents move icebergs, they slowly melt, releasing stones of all sizes, picked up by the glaciers from which they calved. The stones rain down through the ocean water and embed themselves in the soft, smooth mud that covers most of the deep seafloor. Dropstones mean icebergs, and icebergs require tongues of glacial ice extending to sea level and calving into the ocean.

Larsen homed in on the times of dropstone deposition by dating single-celled, microscopic fossils of ocean-dwelling plankton, called foraminifera, that were present in the muddy sediments. In his paper, Larsen revealed that southeastern Greenland got its ice 7 million years ago, around the time that the first member of the hominid lineage stood up and started walking on two feet.[38] By 2.6 million years ago, the start of the Pleistocene Epoch, the rain of dropstones had become nearly continuous as Earth cooled. Between then and now, long, cold glacial periods, punctuated by brief but warm interglacials, came and went about fifty times. Larsen showed that Greenland had a much longer glacial history than any geologic evidence on the island had previously suggested.

The *JOIDES Resolution* did not collect cores exclusively for climate scientists. In 2012, a consortium of eight oil companies hired the ship to collect a core in Baffin Bay, off the northwest coast of Greenland.[39] The oil companies didn't care about the mud, sand, and rocks left by millions of years of glaciers coming and going—they were looking for oil and gas in the rock below. So after the cruise ended, they handed the upper couple hundred feet of one core to the Geological Survey of Denmark and Greenland, known better as GEUS. They also provided the incredibly detailed seismic data used to decide where to collect the core. Interpreting these suboceanic squiggles allowed Danish geologist Paul Knutz to craft a story of Greenland's ice sheet as it came and went over the past few million years.[40]

I first met Knutz in 2018 when he invited me to Copenhagen for a meeting at GEUS. He had assembled a group of scientists to refine a proposal for drilling a new set of ocean sediment cores that would provide a better understanding of how Greenland and its ice sheet behaved over the past 15 million years. Knutz was trying to look back into the Miocene, an epoch that started with warming about 23 million years ago and then cooled until its end about 5.3 million years ago. He planned to collect cores off the west coast of Greenland. They would test whether the evidence of Miocene ice that Larsen had found on the other side of the island covered all of Greenland or just the mountains of East Greenland.

I was in Copenhagen as a dating expert, or in the jargon of earth science, a geochronologist. Since the early 1990s, I have measured rare isotopes of beryllium, aluminum, chlorine, and carbon found in earth materials. Cosmic rays, high-energy protons and neutrons from outside our solar system that continually bombard Earth day and night, create the isotopes I study. On my last afternoon in Denmark, Knutz handed me a half-dozen pounds of brown mud and sand dried almost rock hard. There were five samples—identical three-inch-wide cylinders each sealed in a plastic sleeve. Only the core depths, written in permanent marker, differentiated them. This material was till, a sediment produced at the bottoms of glaciers and ice sheets, consisting of a mixture of whatever the ice ground up and moved around. Much of Canada, northern Europe, and the New England and Midwest regions of the United States, as well as the margins of Greenland, are home to deposits of till. Once an ice sheet is gone, the presence of till tells geologists that glaciers once covered the area—the challenge is dating that till.

Knutz warned me to handle the samples with care because he thought they were about 2 million years old. Having flown to Copenhagen with only a backpack, I rushed downtown before the stores closed to buy a hard-sided suitcase small enough so I could carry the samples on the plane with me. The next morning, as I headed to the airport to fly home, I wasn't sure that the five bags of sediment would make it through airport security. When I arrived in Montreal, Canadian Customs nicely waved me through. But the U.S. agent who greeted me when I crossed into Vermont by car was more suspicious. He didn't know what to make of 2-million-year-old mud. Was I trying to import soil from abroad, strictly forbidden contraband? After a few phone calls, he decided that the till was rock, not soil, and that there was no need for its confiscation. None of us knew it at the time, but analyzing these brown, gritty, desiccated cylinders would be our dry run for analyzing Camp Century's frozen soil.

In Vermont, I handed the samples off to Drew Christ, another geologist on our team. As geochronologists, we like sand. We can separate it, wash it, poke it, prod it, boil it in acid, and if need be, dissolve it. Most marine sediment cores are mud and thus much more difficult for us to work with. Within a few weeks, Christ started processing the core, soaking the hardened chunks in deionized water and using an ultrasonic bath to help break them up. Then, he used a stack of brass sieves and more pure water to separate the material by grain size. This core was unusual. Not only were the samples large, but they were chock full of sand and gravel along with the mud. We assembled a nine-person team, a conglomerate of geologists, physicists, geochemists, and geochronologists, to analyze the till. Our goal was to understand what was happening on Greenland through this material that the ice sheet, which had scoured the land surface beneath it, left behind offshore.

Less than a year after I flew back from Denmark, we published our results in a paper titled simply "The Northwestern Greenland Ice Sheet during the Early Pleistocene Was Similar to Today." In it, we argued the case that the Greenland Ice Sheet of 2 million years ago behaved much as it does right now.[41] It bulldozed plants along its edges that looked like today's plants, and it had already eroded meters of rock—a lot of rock, but not enough to ream out deep fiords. But our most important finding was that till, at which many geochemists look askance, holds lots of ice-sheet secrets. And if we applied the right methods, it would share those secrets.

* * *

TODAY, FROZEN WATER covers most of Greenland, even at the height of summer. Most of that ice is part of Greenland's ice sheet, which is second in area and volume only to Antarctica's. Seventy years after the first ice-thickness measurement at Eismitte, airborne radar replaced explosives as the best way of peering inside ice sheets. Operation IceBridge, a project of the National Aeronautics and Space Administration (NASA), crisscrossed Greenland from 2009 to 2019, taking millions of ice-thickness measurements from airplanes. The data ultimately revealed that the ice is thickest at the center of the island and thins gradually, but not smoothly, toward the edges.

The radar data show that Greenland's ice sheet holds more than 700,000 cubic miles of frozen water. That's 217 times more water than fills Lake Superior, the third largest lake in the world. If the ice sheet were to melt completely, the resulting water could fill 1.17 trillion Olympic swimming pools. Melt Greenland's ice, and sea level around the world would go up, on average, about twenty-four feet.[42] Many hundreds of millions of people around the world would then find their homes flooded.[43] When the ice is gone, what we consider coastal real estate today would become ocean-bottom property.

Greenland's ice sheet is so massive, in fact, that the earth below it sags in places. The depression reflects the structure of our planet, with its thin, rigid crust overlying a solid but softer mantle. Imagine a not quite fully inflated beach ball; if you were to make a fist and slowly push your hand into the ball, the vinyl "crust" beneath your fingers would sink as the airy "mantle" flowed away. Around the edges of your fist, you'd see a diffuse ring (the forebulge) where the ball bulges out ever so slightly to accommodate the volume of air pushed out of the way by your fist. Ice sheets have a similar effect on Earth's surface. As they grow larger, the softer mantle creeps slowly outward, the solid crust below the ice sags, and the land surface falls. When an ice sheet melts, the mantle flows back under the crust, and the land surface rises. These processes are known as isostatic depression and isostatic rebound, respectively. Both matter to coastal communities, whether or not they are close to or far away from an ice sheet.

At the center of Greenland, the ice is 10,000 feet thick, and the land surface is 2,700 feet lower than it was when the ice was gone.[44] At the edge of the ice sheet, where the ice is thinner, the depression is less—imagine moving away from a large person sleeping in the center of a saggy mattress. The dependence of isostatic rebound on ice thickness explains some seemingly odd field observations, such as ancient shorelines and beaches, once flat-lying markers of past water levels, that now rise toward the center of a vanished ice sheet.

The radar data also reveal what's hiding under Greenland's ice. The island is a bowl. Rugged mountain ranges ring the edges. The center of the island is thousands of feet lower than the margins, and the topography is subdued—except for the Grand Canyon of Greenland, a channel cut 2,600 feet into rock that runs 470 miles from the center of the island toward the northwest. Today, the canyon hides beneath thousands of feet of ice. Well before ice covered Greenland, a river, perhaps an Arctic version of the Colorado, cut the canyon. It's not the only evidence of water moving across Greenland's landscape when it didn't have any ice. Under the ice, there's a thirty-by-ninety-mile lake basin filled with sediment, and there are sub-ice channels more than a hundred miles long that coalesce downstream toward the coast—the hallmark of another former river network now buried under ice.[45]

Near where Greenland's Grand Canyon empties into Baffin Bay, the ice-penetrating radar reveals a nearly perfect circle, nineteen miles wide and 1,000 feet deep, larger than the city of Paris. It's an impact crater, named Hiawatha after the glacier that covers it.[46] Analysis of minerals, damaged by the impact and now washing out of the ice in meltwater streams, shows that about 58 million years ago, an iron meteorite slammed into Greenland.[47] Although it was large, this impactor was seven times smaller than the rock that struck the Yucatán Peninsula 8 million years before. That Central

American impact ended the reign of the dinosaurs and plunged Earth into a dark, years-long winter as soot from wildfires blocked the sun and shut down most photosynthesis.[48] There's no evidence that the Hiawatha impact affected global climate significantly.

If Greenland's ice vanished tomorrow, the resulting landscape would be unrecognizable. The ocean would pour into the center of the island, leaving an archipelago around its edges. Today's mountains would be tomorrow's islands, and the center of the ice sheet would become an inland sea. But the arrangement wouldn't last. Isostasy would kick in, and over a few thousand years, the land would rise out of the ocean, rapidly at first and then more slowly. Eventually, most of Greenland would again be above the waves.

Understanding the properties of ice is key to predicting its behavior. Ice is an unusual solid because it flows if the conditions are right. Hit it with a hammer, and it shatters—that's brittle failure. But hang a piano wire over a block of ice with weights on both ends and the wire will slowly "cut" its way through the ice, leaving no hole in its wake. This trick is plastic behavior, which means that if you don't hurry ice, and if that ice is sufficiently stressed, it will flow without cracking or shattering.

In 1958, English physicist John W. Glen published a simple equation that described how quickly ice flows (the strain rate, which he called e):

$$e = (\sigma/A)^n$$

Glen's equation includes three values. The Greek letter sigma, σ, is the amount of stress on the ice—calculated by measuring ice thickness and the density of the ice, and using the well-known value for the acceleration of gravity. The other two values (A and n) are determined in the cold lab by deforming ice under pressure using machines. A is the rate factor, thought to represent the effect of temperature, the size and shape of ice grains, and impurities in the ice, including water between the grains. The exponent, *n*, describes how quickly the strain rate goes up as the stress increases.

Most glaciologists use a value of about 3 for *n*, but where stresses and strain rates are high, 4 seems to be better.[49] *A* varies with temperature, increasing sevenfold as ice warms from 14°F to the freezing point. That means that warm ice deforms more quickly than cold ice, and that once ice starts to move quickly, it will gain speed. The former property matters when people living in the ice warm their surroundings—for instance, by heating the living space at Eismitte. The latter quickens the flow of outlet glaciers—those that pass through narrow fiords as they move away from the ice sheet toward the ice margin and the sea. All glaciers move as their ice deforms internally following Glen's flow law.

Ice has another unusual property: it melts under pressure. Beneath 7,000 feet of ice, the pressure is nearly 200 times greater than atmospheric pressure at sea level. As a result, water freezes at just over 30°F. That odd behavior means that at the bottom of a massive ice sheet, there's often liquid water, even though the temperature is several degrees below what we consider the freezing point of water. That liquid water, which saturates the rock and sediment below it, can help the glacier flow by sliding. Such sliding glaciers usually move many times faster than the deformation of ice according to Glen's flow law would suggest.

These wet, sliding glaciers are "warm-based," indicating that the glacier is melting at its base rather than frozen to the material below. Sliding happens only if liquid water is present at the base of the ice. Warm-based glaciers are effective agents of erosion, stripping weathered rock from the land surface and shaping the bedrock below into forms elongated to match the direction of ice flow. Warm-based glaciers smoothed and rounded the rock outcrops on which Greenland's cities and towns perch. That ice excavated U-shaped glacial valleys and deepened the fiords that characterize the island's coastlines. Warm-based ice leaves behind striations: long, linear gouges in rock at the bed of the glacier.

For glaciers frozen to the rock and sediment below them, slow deformation of the ice is the only way they can move. These "cold-based" glaciers can do some strange things. In northern Greenland, I've hiked and helicoptered over terrain where the bedrock, shattered by frost and weathered by water, was so soft that you could dig into it with a shovel—even though from the air, it looked like a solid outcropping of rock. Yet we knew that the ice sheet had once been there, because scattered on this rotten rock were erratic boulders the size of pickup trucks—rocks of a composition different from that of the bedrock that had been carried in from elsewhere. Cold-based ice, frozen hard to the weathered rock beneath and moving by deformation only, delivered these hard, fresh boulders. When the ice melted away, the boulders stayed behind.

Some people call such a cold-based glacier a "ghost glacier" because it leaves no trace of flowing once it vanishes—no rounded outcrops and no striations.[50] Ghost glaciers, while they linger on the landscape, do an excellent job of protecting what's beneath them even if it's fragile. On Baffin Island, University of Colorado geologist Gifford Miller and students working with him have spent the last decade collecting perfectly preserved plants from the edges of small cold-based glaciers that are melting away in our century's heat wave. The recent reappearance of this ancient, long-frozen vegetation suggests that Earth is warmer now than it's been since the ice buried the ecosystem.[51] Radiocarbon dating shows unambiguously that ice covered and killed these plants more than 40,000 years ago.

Understanding which parts of the Greenland Ice Sheet are cold-based, and which parts are warm-based, is critical to predicting how the ice sheet moves and changes over time. It's also critical for knowing where frozen sediment, and thus a record of Greenland's past, remains preserved under the ice. It's not simple to determine where the bed is frozen and where it's not without drilling a hole, but remote sensing from airplanes and satellites provides some reasonable clues. Between 2003 and 2010, NASA's ICESat mission measured the elevation of millions of points scattered across Greenland's ice sheet. When NASA retired the satellite, they replaced it with IceBridge. Using ice-penetrating radar allowed scientists to peer inside the ice sheet and to see the bed below.

NASA's Joe MacGregor used four quite different kinds of information to make a best-guess map of Greenland's warm-based and cold-based ice.[52] He synthesized computer models, layers in the ice sheet identified by radar, the speed at which the ice was moving (did it exceed the Glen's law, deformation-only "speed limit"?), and ice-surface topography, because rough spots indicate warm-based ice.

By MacGregor's analysis, 43 percent of the Greenland Ice Sheet is warm-based, and 24 percent is cold-based. The other 33 percent remains uncertain. His map shows most of the cold-based ice below the center of the ice sheet and to the north. Almost all the ice near the coast is warm-based. MacGregor is quoted as saying, "I call this [paper] the piñata, because it's a first assessment that is bound to get beat up by other groups as techniques improve or new data are introduced. . . . But that still makes our effort essential, because prior to our study, [they] had little to pick on." There is little ground truth. Fewer than twenty boreholes have reached the bottom of the Greenland Ice Sheet, and only a half dozen of these are away from the coast.[53] Some encountered liquid water. Others found ice frozen to rock or soil. Every new borehole through the ice will test MacGregor's map.

Glaciers are a fickle, frozen scrapbook. Ice, along with the sand and gravel it both moves and preserves, is the keeper of Greenland's history reaching back tens, thousands, and even millions of years. Because much of Greenland's ice sheet is moving, flowing, and eroding, it can be tough to find old ice, soil, and rock. But old materials are key. They have the potential to reveal the history of Greenland through warm times and cold, and thus the history of our changing planet and its climate. Predicting our collective future in this time of rapidly changing climate requires understanding Greenland's history over the last several million years. One key to understanding that history is analyzing sediment preserved below the ice sheet—especially where that sediment is frozen to the rock below and the ice above.

* * *

The surface of Greenland's ice sheet away from the margin is a nearly featureless plain. In the interior, the only topography is layers of snow sculpted by the wind into *sastrugi* (Russian for furrows)—elongated troughs and ridges that can be meters high. Toward the margin of the ice sheet, the icescape grows more extreme and treacherous. Crevasses form where the ice pulls apart as it moves over bumps on the bedrock below. These deep chasms in the ice are easy to avoid in the summer, when you can see them, but when shrouded in winter snow, they swallow men and machines as the fragile powder draped over them gives way. The edge of the ice sheet is steep. Early explorers struggled to get their gear onto the ice using cables, winches, horses, and dog teams.

Most of Greenland is shades of bright white and light gray; today, only the ice-free margins are darker green and brown. In the winter, the snow-covered ice sheet glistens. In the spring, as the sun rises higher in the sky and the air warms, the winter's snow begins to melt—first in the south and along the coast where the air is warmer. Slowly, the melt creeps inland and up the ice sheet to higher elevations. When snow melts, it goes gray as water fills spaces between the rapidly changing snow grains. A feedback loop sets in: wet snow reflects less sunlight and absorbs more solar energy than dry snow. Melt encourages more melt.

Once the past winter's snow is gone, older gray ice appears, and rivers of water, derived from melting snow and ice, flow across the glacier. The meltwater etches meandering channels into the ice. Some of it flows directly to the sea; some takes a detour to watercolor-blue ephemeral lakes that dot the edges of the ice sheet in the summer. The lakes have an unpredictable habit of draining catastrophically into cracks in the ice. Their waters tumble hundreds, even thousands of feet into moulins, vertical holes in the ice, which funnel meltwater to a network of channels running along the base of the ice sheet, all confined in tunnels of ice.

At the ice margin, the meltwater emerges in roiling outwash streams that can be a hundred or more feet across. I've stood by one of these streams and felt the ground shake as boulders the size of basketballs bounce and roll past. The streams are gray-green with glacial silt, so turbid that you hear the rocks moving but rarely see them. As summer wanes, the days shorten and a chill settles in. The melt slows, and flow in the outwash streams wanes. Soon, the first snows of fall blanket the ice sheet, and the yearly cycle starts again.

Driven by gravity, ice flows from the center of the ice sheet to the edges and closer to the coast. Everything entombed in the ice also heads seaward—sand, boulders, and the empty cans and human waste left behind at Eismitte. In much of Greenland, ice moves slowly, just a few tens of feet per year. At ice divides—the high points of the ice sheet and the best places to drill ice cores—the flow is even slower. In the deep fiords that ring the island, however, the flow is quick enough (up to a dozen miles per year) that you can see and hear the ice move; it's hard to look away. Where the ice is warm-based, flow is faster in the summer, when meltwater lubricates the bed, and slower in the winter, when it does not.

Like people, ice sheets grow and shrink over time. Early explorers trekking for months across the Greenland Ice Sheet lost twenty or even thirty pounds. Geologists eating slices of cheesy musk ox pizza from the shop next to the Kangerlussuaq airport gain weight. The ice sheet gains mass mostly from snowfall, but also when rain falling on cold snow freezes—a decidedly uncomfortable experience in the field. The ice sheet loses mass in two ways. First, fast-moving outlet glaciers reach the ocean, where massive towers of ice calve off the steep ice front, birthing icebergs. Of equal importance is surface melt driven by the sun's energy, both directly, as it beats down on the ice twenty-four seven in summer, and indirectly, as warm, moist winds melt the ice.[54] Mass balance—the inputs of ice and snow versus their outputs—is a measure of glacier health and, eventually, survival.

Elevation, because it largely controls air temperature, determines the local mass balance. Melt dominates the low, warm places on the ice sheet, those near Greenland's coast. On average, these areas lose more mass than they gain and are the ablation zones. High, cold places deep in the interior of Greenland, like Eismitte, rarely experience melt or rain—only snowfall. These places are the accumulation zones. Between the two, there's a mathematical line of demarcation, a place known to glaciologists as the equilibrium line altitude (ELA). That's the elevation where, on average, ice gained equals ice lost.

It's easy to visualize the ELA if you fly over Greenland in September, as melt season ends. You'll see a boundary on the ice sheet, often sharp, sometimes not. Above the boundary, where the previous winter's snow survived the summer melt, the ice surface is white. Below it, where the past season's snow is gone, the surface is a dull gray. The gray is firn, old snow transformed by melting, freezing, and melting again. Its grains are darker, larger, and rounder than new snow and thus easily distinguished on the ground and from the air. Today, satellites map Greenland's firn at the end of every summer just before new snow starts to fall.[55] The average elevation of Greenland's firn line is 4,500 feet; it's higher in the south where the climate is warmer, lower in the north where it's cooler.[56] As the climate cools, the ELA and the firn line fall. As the climate warms, they rise, and the area of the ice sheet that's gaining mass shrinks.

Understanding how Greenland's ice sheet, its ELA, and its mass balance responded to prior, naturally driven warmings is critical to predicting what will happen next as our climate warms. That's where ice cores, and the sediment frozen beneath the ice sheet, come in handy.

The Snowblast pulverizes snow on the Greenland Ice Sheet, 1958 or 1959. The result is a stiffer, stronger runway. *U.S. Army photograph.*

2

FROZEN WATER

Snowflakes fall to Earth and leave a message.

—HENRI BADER, *The History of Early Polar Ice Cores*

Snow is exquisitely frozen water. Every marvelous crystal is unique. It's fragile, intriguing stuff that falls from the sky in all sorts of shapes and sizes. Snow also carries memories of when and where it crystallized high in the atmosphere. The frozen molecules of water record, in the isotopic composition of each of their oxygen and hydrogen atoms, how cold the air was when they froze and where the air came from. At birth, snowflakes wrap themselves around tiny bits of earth and sea suspended in the atmosphere. Locked to these condensation nuclei, they begin a trip from the clouds to Earth's surface. If a snowflake lands on an ice sheet, its journey back to the ocean and the atmosphere could take thousands, even millions, of years.

Wilson Alwyn Bentley invented the study of snow in Jericho, Vermont, just ten miles and a few towns away from the university where I work. Known now as Snowflake Bentley, he spent years capturing, observing, and categorizing Vermont snow as it fell.[1] In 1885, at age nineteen, he began photographing snowflakes. A homeschooled, self-taught farmer, Bentley created the equipment and procedures needed to capture these delicate, tiny forms on film without their melting. Each had a different shape, but he found that he could sort those shapes into discrete categories. Most flakes were hexagons. Some were flat plates. Others were columns. Dendrites were lacy. Graupel, coated with rime ice from a moist trip through the atmosphere, was lumpy and irregular. In just a few decades, with no formal science training, Bentley wrote sixty papers and articles about snow. He published a book containing almost 2,500 photographs of snow crystals.

By the time Bentley died of pneumonia in December 1931, he had photographed over 5,000 different snowflakes, and had developed the first classification of snowflake shapes and forms. A hand-drawn poster portraying more than 400 of his snowflake images, arranged by temperature, hangs on my living room wall. The structure of each flake depends on the atmosphere's behavior and the humidity of the air. In 1947, 12,000 of Bentley's snowflake negatives and glass plates arrived in Buffalo, New York, acquired by the city's Museum of Science—appropriate for a city that, on average, gets more than seven feet of snow each year.[2]

With snow, change is inevitable. Snow isn't a stable material because temperatures at Earth's surface, even in frigid Greenland, remain close to the melting point of water. Wind jostles the flakes as they fall, breaking off their lacy spikes. The resulting windblown snow quickly grows dense and strong; on steep, snow-covered slopes, it births slab avalanches. On an ice sheet, windblown snow appears solid and resilient, but it's not. Flakes rapidly metamorphose and lose their original identity. In the snowpack, water molecules, one by one, leave the points of snow crystals and migrate to the concavities between. The icy grains bond to their neighbors. When snow warms to the freezing point, melting takes the edges of snow crystals first, rounding the flakes until they look like an assortment of small ball bearings. As soon as the ice between snow grains warms, it melts, becomes liquid water, and the snowpack loses strength.

Four of Wilson Bentley's photographs from the early 1900s, showing the intricacy and variety of newly fallen snowflakes. *Images property of the Jericho Historical Society, Jericho, Vermont.*

Snowflakes, no matter how they start, eventually transform into bigger, bulkier, and rounder grains with less surface area and more volume.

Snow is a dynamic material, prone to rapid, and sometimes unpredictable, fits of change. That bad habit can make the changes in snow a matter of life or death for people working on, around, and inside ice sheets. For those moving over snow, the character and strength of the flakes—what the military refers to as the snow's trafficability—are critical. Deep, fresh snow can be very soft and thus has little strength making it nearly impassable because it can't bear a load. Men, dogsleds, and snow machines get bogged down because the snow deforms under their weight. Airplanes snag their skis when landing and can flip over. Dense layers of snow often erode into hard ridges, making travel painfully slow and difficult for both people and their machines. Melting snow, saturated with water, is loose and sticky, slowing travel and stressing mechanical equipment.

When a snowflake lands in the accumulation zone of an ice sheet, it is buried by the next snowflake and the next and the next. Over time (it might be weeks, it might be decades), snowflakes transform into the icy firn that Sorge sampled at Eismitte. Pressure from the snow above compresses the snow below and squeezes out much, but not quite all, of the air. Over time, firn grains merge, finishing their transformation into solid glacial ice. Where snowfall is heavy and warm, the transition takes a decade or two. Where snowfall is less and the climate is cold, making glacial ice from snow can take centuries. The resulting ice retains much of the chemical and isotopic character of the snow, and then the firn, that birthed it, even though the ice—a solid, slightly blue mass—looks nothing like the snow from which it formed.

IN MAY 1945, Nazi Germany collapsed. Three weeks later, Henri Bader, a Swiss geologist with expertise in snow and a penchant for numbers, boarded a plane to Miami with his wife, Adele.[3] By fall, he was working at Rutgers University in New Jersey, teaching classes and advising students.[4] The U.S. military quickly took an interest in Bader due to his expertise in snow and ice. Before the war, Bader had worked at the Snow and Avalanche Research Commission of Switzerland. He turned to snow after eight years of university research in Zurich using X-rays to probe the atom-by-atom structure of crystals in rocks. Ice is crystallized water, so this was a manageable intellectual leap.

In the summer of 1938, Bader had left snow and Switzerland for Argentina.[5] He sailed on MV *Piriapolis*, a Belgian vessel that repeatedly carried Jews from Europe to Argentina. Its passengers included some of the world's best chess players, who were competing in the last international tournament before the war.[6] At least nineteen of these players, all Jews, failed to return to Europe after the tournament.[7] A month after he landed, Bader married Adele Christen, who was also a Swiss expat living in Argentina.[8] She spoke German and English. He spoke those and French to boot. Together, they learned Spanish.[9] On September 1, 1939, Germany invaded Poland. Soon after, Luftwaffe bombers sank the *Piriapolis* off Le Havre as France fell to the Germans.[10]

The Baders stayed in the Southern Hemisphere for the rest of the war. They prospected in Argentina, moved to Colombia, and settled in Curaçao, where Bader spent the rest of the war as a quarry superintendent.[11] The quarry mined phosphorus ore known for its exceptional purity even though it got its start as million-year-old bird guano.[12] Phosphorus was and is a strategic element, used for bombs, smoke grenades, and fertilizer. In 1942, German U-boats began to attack oil tankers in the West Indies, and over a thousand American soldiers arrived to defend the island of Curaçao and its refineries because 70 percent of the Allied forces' fuel came from the West Indies.[13] Bader wasn't far from the American military, even before he came to the United States.

Bader and his science thrived on Cold War military dollars. By spring 1947, the War Department General Staff sent the multilingual Bader to Europe, where he met with scientists in England, France, Switzerland, and Germany and assessed what each knew about snow and ice mechanics.[14] His European reconnaissance was both the prelude to and the basis for a Pentagon conference about snow, ice, and the military's needs. That meeting, on August 12, 1947, birthed the U.S. Arctic science program, which stressed basic research in support of national security—both of which required understanding the cryosphere, the frozen portion of Earth, and how it changed over time. By 1949, Bader applied for American citizenship and moved to the University of Minnesota. There, he worked alongside B. Lyle Hansen, who had run the Sierra Snow Laboratory in California and would soon go on to lead the Army's ice-coring efforts in Greenland. The U.S. Army Corps of Engineers supported both men and their science.

In 1949, Bader laid out a research agenda that rejected specialization and embraced the combined power of geology, chemistry, biology, physics, and engineering to do cold-region science. Bader emphasized that

> the rapid specialization of scientists near the turn of the century was disastrous to glaciology. It was classified as a geological science, and from then on glaciers were studied very largely by geologists, with the emphasis on glacial geology. Petrographers and mineralogists generally failed to consider snow and ice as sedimentary and metamorphic rocks and paid them little attention. Physicists were not particularly attracted to the study of a material requiring a refrigerated laboratory, and contented themselves with routine measurements of the more common physical properties of ice. Chemists and biologists saw little to be gained in the study of practically pure and sterile ice. And, finally the remoteness of glaciers from centers of industry and main lines of communication effectively prevented them from becoming an object of engineering research.[15]

Bader's agenda would shape polar science for the next two decades. His visionary approach to gathering fundamental knowledge about ice and snow behavior, specifically by crossing disciplinary boundaries, catalyzed an entire field of both applied and academic science and engineering.

While Bader and the U.S. military approached their work differently, each benefited from the other's work. The Army wanted its men and equipment to function on a frozen battlefield. Bader wanted to know how the cryosphere worked. The Army formalized their collaboration in 1949 when it established the Snow, Ice, and Permafrost Research Establishment (SIPRE), in which Bader rose through the ranks to become chief scientist. During the 1950s, Bader, the people with whom he worked, and the military built the scientific foundation for snow studies and ice-core analysis.[16]

IN JULY 1950, ski-equipped Air Force C-47s delivered seven tons of drilling equipment to Southeast Alaska's Juneau Icefield—a first.[17] Bader embarked on the expedition that would use that equipment, accompanied by two assistants—Gerry Wasserburg and Anders Anderson—with quite different, but critical, skill sets. The expedition had two goals: first, to figure out how to drill into a glacier, and second, to recover the first American ice core.[18] Their drill camp on the Taku Glacier was sixteen miles from the ocean.[19] The glacier extends thirty-six miles from its inland origin to the coast in a deep bedrock trough hemmed in by rocky mountain peaks. The Taku is almost 5,000 feet thick in places—a record for alpine ice. Seismic data gathered the year before showed almost 900 feet of ice covering the bedrock where Bader planned to drill.

The drilling was part of the fledgling Juneau Icefield Research Project (JIRP), which has surveyed its namesake ice field every summer since 1948.[20] In the late 1940s, the military used these ice-field expeditions to refine the logistics of working on and around ice and snow. Maynard Malcolm Miller, a geologist who ran JIRP for decades, was looking for something quite different: evidence of climate change. As he wrote of a 1949 expedition in Alaska, "Tremendous recession of ice is going on; the Earth definitely seems to be warming up. These glaciers are telling a story. Only when we learn to read it can we know whether the Earth is warming up on a major scale."[21] Miller was well ahead of his time when he worried about melting ice and shrinking glaciers.

Wasserburg came to Juneau because he was a student of Bader's

Henri Bader measuring a crystal of ice during a press conference, February 25, 1953, in the walk-in freezer of the Snow, Ice, and Permafrost Research Establishment, Wilmette, Illinois. *Photograph by Ralph Walters, ST-17623525, Chicago Sun-Times collection, Chicago History Museum.*

at Rutgers. He had grown up poor in New Brunswick, New Jersey. His high school grades were so low that he came to college, after a stint in the military at the end of World War II, only as a night student. Wasserburg was also Jewish. Early on, his dreams met up with religious prejudice. He recalls telling a professor, "Well, I'm going to be a geologist. I'll work for an oil company." The professor said, "Oil companies don't hire Jews, Wasserburg. Let me explain that to you."[22] Wasserburg would go on to become a Caltech professor and America's most influential isotope geochemist, studying moon rocks and pioneering the clean-room chemistry approaches that we and others use today to date the coming and going of Greenland's ice sheet.[23]

Anders K. Anderson, "an experienced diamond driller" from Minnesota, was in charge of the drilling.[24] He and the Pioneer Straight Line Drill Rig, with its three-legged metal derrick that towered above the camp, were on loan from Minnesota's Longyear Company—a hard-rock drilling operation that got its start in 1890 when its founder drilled his first hole in Minnesota's Iron Range. The rotary drilling rig Anderson ran on the ice was a wireline design with a four-cylinder gas engine that drove both the drill and the water and hydraulic pumps associated with it. This meant that each time the drill advanced a set distance (often five or ten feet), the drillers used a wire to bring a tube, holding the core segment, to the surface. Because there was no need to remove a solid metal drill stem and bit, wireline was the most efficient way to pull core segments from deep boreholes in ice or rock.

But these men were not the first to drill holes into a glacier. In 1841, Louis Agassiz, considered by many the father of glaciology, drilled into the Unteraargletscher, a glacier in the Swiss Bernese Alps.[25] He used a chisel-tipped iron rod, to which he kept adding extensions until he ran out. His team's persistence was rewarded with a hole 165 feet deep, but they failed to reach the bottom of the ice. Agassiz came back the next year with more men, a cable, and a heavy iron lance at the bottom. For weeks they raised and lowered the lance, chipping away at the ice. This time, they cut a 200-foot-deep hole before giving up, having still not gotten through the ice.

Later in life, Agassiz was a professor at Harvard and had an outsized impact on studies of geology and natural history. He popularized the idea that Earth had undergone ice ages, times when glaciers and ice sheets had grown much larger than they are today and shaped many of the world's landscapes. In a talk before the Geological Society of London, a hotbed of geology in the 1840s, Agassiz used Greenland's ice as a modern analog for his imagined but long-vanished ice sheets elsewhere. But he was also an avowed white supremacist who claimed Black and white people were of different species. During his long career, contemporaries repeatedly accused him of taking his theories of ice ages from others without giving proper credit—an accusation modern historians support.[26] It's fitting that his tombstone is a glacial erratic—a piece of granite carried south to Boston by the now-vanished Laurentide Ice Sheet, which covered much of eastern and central North America 25,000 years ago.[27]

Since Agassiz's time, people have gotten better at drilling holes in the ground. Most are drilled to tap groundwater for drinking or for industry use. Others recover natural gas or oil, capture geothermal heat, or allow for underground waste disposal. Drilling into glacial ice still isn't a common feat, and the Taku Glacier site, which was wet and warm, made ice drilling all the more challenging. The drill site was in the glacier's ablation zone at a low elevation, 3,640 feet above sea level, and in a maritime climate. Most days were above freezing in July, a month during which, on average, several inches of rain fell. When the team arrived, the ablation season was in full swing, and the snow and ice below their feet melted rapidly. Photographs taken soon after drilling began show crevasses opening between the canvas tents in which the men lived. The drilling platform tilted as the snow beneath it melted. At least the team was not short of water with which to flush the icy drill cuttings out of the borehole.

There was no established procedure for drilling or analyzing ice cores—the men on the Taku Glacier figured it out as they went along. First, they dug a long pit in the glacier, which became their makeshift cold laboratory and core storage area because the temperature in the ice didn't go above freezing. A small oval doorway led to this science trench, which had a white tarp over the top and a short wooden ladder leading down from the ice surface.[28] Then, they

In the summer of 1950, as snow in the ablation zone melted, crevasses began to open near the tents at the Taku Glacier drill camp. *Photograph by Maynard Miller, Juneau Icefield Research Program, provided by the Maynard & Joan Miller Family.*

drilled three holes into the glacier. Even the deepest hole, which took fifteen days of drilling to reach a depth of 300 feet, didn't hit the bottom of the ice. The cores were narrow at only two inches, a diameter so small that drilling often crushed the ice.[29] When they succeeded in retrieving a core segment, several men would remove the clear but layered ice from the core head and place it gently into a wooden storage box.

Bader and Wasserburg studied the deeper half of the core. In the science trench, they cut the core into pieces, making thin sections of the ice. Staring through microscopes, the men analyzed hundreds of ice crystals from the thin sections.[30] For three weeks, they sketched the crystals' shapes and measured their angles. By shining polarized light through the ice, they determined the orientation of each crystal. Their research opened up a new field of inquiry into how glacial ice deformed at a microscopic scale. This was more than academic science; it was critical knowledge if you wanted to drive a tractor across a glacier, dig tunnels in an ice sheet, or land a cargo plane on an ice runway.

When the drilling finished, the men lowered 250 feet of hollow aluminum pipe into the deepest hole; they would use the soft metal as a proxy to measure how the Taku Glacier deformed over time. Data came from a well-logging tool equipped with a camera, compass, and plumb bob that hung straight down. The tool allowed them to measure the direction and magnitude of the pipe's tilt at different depths. For several years, JIRP participants returned and remeasured the aluminum pipe as it recorded ice movement and deformation in three dimensions over time. In another hole, they installed a string of thermistors, which measured ice temperature as the seasons changed. In the center of camp, they put air-temperature sensors on a spindly metal tower that was thirty-five feet tall.[31] On the Taku Glacier, Bader was leading a new, data-driven approach to understanding the behavior of ice and snow. His 1949 vision—a systemic understanding of the inner workings of snow, ice, and glaciers—was starting to take shape, one field site at a time.[32]

In the summer of 1950, no one knew that the Taku Glacier was an oddball. Most of the world's glaciers started retreating in the 1970s, but the Taku continued to advance for decades, even as the climate warmed and its peers shrank.[33] That behavior has now come to an end. The U.S. National Oceanographic and Atmospheric Administration (NOAA) and its satellites have been watching the Taku's shrinking ice from the sky. In 2019, NOAA reported that the

Taku had finally joined the rest of the world's glaciers and started to retreat. In the words of Mauri Pelto, a glaciologist who studied the Taku for decades, "This is a big deal for me because I had this one glacier I could hold on to. But not anymore. This makes the score climate change: 250 and alpine glaciers: 0."[34]

Bader and Wasserburg's expedition proved that ice coring was possible, even if core recovery was inconsistent and the quality of samples varied.[35] Still, the first ice core was good enough that samples of the ice went to the University of Chicago for analysis of oxygen isotopes, and to other labs that measured concentrations of trace elements.[36] After the Taku, Wasserburg left glaciers behind to solve the mysteries of the universe and the solid part of Earth, but he never forgot the Baders, visiting them whenever he could.[37] Bader would now become the central figure in the American military-industrial Arctic science complex, the focus of which was shifting quickly to Greenland.

After World War II, the Army Corps of Engineers began building runways and camps on the snow of Greenland's ice sheet. First, there was Project Snowman (1947), then Project Overheat (1950), and finally, in 1953, Project Mint Julep.[38] These projects built on one another and on the work done during the war in East Greenland. Declassified reports list the players: the Soils Laboratory, the Arctic Construction and Frost Effects Laboratory, the Airfields Division, Military Construction, the U.S. Air Force, and the American Geographical Society. The last group, an organization of professional geographers, didn't fit in with the rest.[39]

F. Alton Wade, an American geologist, explains the connection. Wade was a polar veteran who had seen a lot of things go wrong in cold places. He participated in Admiral Richard Byrd's second expedition to Antarctica in 1934, during which Byrd nearly died from carbon monoxide poisoning when the stove he was using failed. Just before World War II, Wade served as chief scientist on Byrd's third expedition. This one carried a fantastical vehicle called the Antarctic Snow Cruiser—a fifty-five-foot-long polar mobile home for five people. It had ten-foot-diameter tires and a diesel-electric propulsion system. While the Snow Cruiser performed admirably as a warm bunk room, and even had a rooftop rack designed to hold a biplane, it was a dismal transportation failure. The smooth tires had no grip and spun in the soft snow. Although designed to reach the South Pole from the coast—its supporters printed and mailed postal covers celebrating this achievement even though it never happened—the cruiser's longest trip was a mere ninety-two miles, all in reverse. It didn't have enough traction to move forward.

After his Antarctic expeditions, Wade had led the Ice Cap Detachment in East Greenland, spending two years near, and on, the ice sheet. At both poles, Wade lived the failures of American military engineering. He dealt with dead Weasels stranded on the ice as transmission after transmission failed. His mechanics nursed finicky and unstable motorized snow sleds back to life. Wade experienced the inadequacy of U.S. military–issued Arctic clothing personally. But successes also shaped his view of the poles—the Weasel expedition to the summit of Greenland's ice sheet; stranded airmen surviving in a snow cave for a winter, supplied only by airdrops; and the daring rescues done by seaplanes belly-landing on the ice, not once but several times.

After the war ended, Wade authored a long paper in the *Geographical Journal* filled with photographs of remote East Greenland. That paper became the outline for 1947's Project Snowman, the U.S. military's postwar foray back to Greenland.[40] In August, a C-47 airplane equipped with skis landed on the ice sheet about ninety miles east of Kangerlussuaq. There, Air Force officers and men from the Army Corps of Engineers set up Project Snowman's camp. Their demonstration that aircraft could reliably land and take off from runways on the ice sheet changed Arctic science dramatically. No longer did scientific expeditions need to rely on sledges, dogs, and men pulling loads over the ice. Machines—specifically, aircraft—would now do much of the work.

Three years later, Project Overheat picked up where Snowman had left off, with more engineering, more airstrip development, more snow testing.[41] Project Mint Julep, the 1953 finale, got into the operational details with a camp and three different test runways over an area of more than fifty square miles. The military's goal was to figure out where on the ice sheet, and when, planes could reliably land and take off. Within the area investigated by Mint Julep, conditions on the ice sheet varied dramatically. At lower elevations and near the coast, in the ablation zone, snow melted away by the height of summer, exposing hard gray ice dotted with lakes and meltwater streams. Farther inland and at higher elevations, in the accumulation zone, the snow stayed mostly dry and there was distinctly less summer melting.

When the weather cooperated, planes could come and go from the surface of the ice sheet. Skis worked on softer snow but could freeze onto the runway when the snow cooled. Wheels worked on ice. Belly landings in soft snow were also an option. But if the snow was too soft or the planes too heavy, there was no taking off. There was a work-around using JATOs, short for Jet Assisted Take Offs.[42] These were strap-on rockets, one-and-done thrusters that could be attached to a plane's wings. They provided enough lift to get the plane off almost any snow surface. Today, the large hollow cannisters of empty JATOs sit perched around the U.S. National Science Foundation dorms in Kangerlussuaq, where I've stayed on most of my trips to Greenland. They enjoy a second use as waist-high ashtrays.

The military engineers tried lots of things to trick snow into behaving. Some worked. Some didn't. They compacted runways using rollers. That worked. Salting a snowy runway made it far stiffer (today this trick firms the snow in competition ski runs), but it took 30,000 pounds of salt to improve a single runway—not a practical solution.[43] The engineers mimicked the effects of wind by grinding snow to a fine powder. It fused together quicky, making for a much firmer runway surface, especially if a heavy roller was passed over the ground snow. They even created a purpose-built machine called the Snowblast, which was a snowblower-tractor chimera.[44] It worked, sort of, but the idea of milling snow to increase its strength turned out to be less critical for runways than when constructing roofs over snow trenches.

The Army continued to learn that the strength of snow changes radically over both time and space because snow is so close to the melting point of ice. Warm things don't get along well with snow. Many of the drab olive-green Mint Julep camp buildings, well heated but not well insulated, settled into the snow. Measurements made that summer showed that the heavier and warmer the buildings, the more they settled. Dealing with the fickle behavior of snow was an inconvenient truth for the military. To better understand snow, in the summer of 1953, the U.S. military began drilling its first ice cores in Greenland.

There's a photograph at the back of the final Mint Julep report titled "Power Drill Rig." The grainy black-and-white image shows a simple metal derrick with a pulley at the top. The rig is about twenty feet tall, and it sits on a pair of wide skis. Hanging vertically just above the snow is a hollow-stem auger, a five-foot section, like the kind used to drill shallow soil borings. Over one summer, that rig drilled thirty-four holes in the ice near eight different snow and ice runways; the deepest extended thirty-three feet into the ice.[45] In the same report are photographs of the resulting ice cores, sliced into round, crystalline slabs a few inches across, the size and shape of small teacup saucers. Other photos show ice and snow cores crushed to determine the load they could withstand before failure.

Mint Julep's final report was an operating manual for building a base on the ice sheet and, in retrospect, a glimpse into the future.[46] The Army's data tracked snow as it cooled and warmed. Hand-drawn graphs documented the transition from snowflakes to firn to glacial ice. Using bright-colored aniline dye, engineers traced summer meltwater as it moved through snow and firn aquifers. Graphs, tables, equations, sketches, and photographs, all of which military engineers needed so they could build and maintain snow runways and camps on the ice, filled the report.

For ice-sheet scientists, the Mint Julep data provided a critical foundation for predicting how snow and ice change and deform over time, and thus for determining how best to collect and interpret ice cores.

The metal tubes of Site 1 soon after installation in 1953. Within two years, snow buried entrances to the tubes, forcing the men to come and go through the hatches. *U.S. Army photograph.*

INTO THE ICE

> It's all there . . . frozen in time. We're just beginning to unlock the secrets in the ice.
>
> —CHESTER LANGWAY, *Los Angeles Times*

In the fourth century BCE, the Greek explorer Pytheas sailed north from what is now France. He and his men circumnavigated the British Isles and kept going, sailing six more days to a place the Greeks had never been. They stopped where "the earth and the sea and all things together are suspended, and this mixture is . . . impassable by foot or ship."[1] They called the place "Thule" and turned for home, their voyage probably stymied by summer sea ice. Only fragments of Pytheas's writings survive, but the name Thule stuck and, for 2,000 years, was used to describe the world's most remote places.[2] The meaning of *Thule* is obscure; suggestions include "isles of darkness," "most remote land," and "the end." Each of these descriptions fits northwestern Greenland, and its northwestern settlement also named Thule, quite well.

Modern-day Thule got its start when Robert Peary, the American who later claimed, some think erroneously, to have been the first to reach the North Pole, spent several months of 1892 there. Recuperating from a broken leg, he remained in a small hut called Red

Stocks of prefabricated construction materials stockpiled by the U.S. Army to build Thule Air Base cover barren permafrost in 1951. *U.S. Army photograph.*

Cliff. Later that year, Peary used dogsleds to cross northern Greenland's ice sheet and returned alive, but hungry. By then, most of his dogs were dead. The starving men had eaten some.[3] Other dogs ate the rest. Two decades later, a pair of Danish explorers, seeking to get away from it all, built and ran a lonely fur-trading outpost at Thule. They interacted with the small population of Polar Inuit who called the area Pituffik. During World War II, the Americans and Danes built a weather station at Thule that remained active after the war. The settlement was 947 miles from the North Pole.

With the coming of the Cold War and the need for a U.S. presence to defend the Arctic against perceived Soviet aggression, Thule grew up. In March 1951, a secret construction campaign, code-named Operation Blue Jay, began. It had the appearance of an

American invasion of the Arctic, as it involved more than a hundred ships loaded with more than 5,000 men and almost 150,000 tons of cargo (mostly construction materials and landing craft packed with trucks and cranes) and GIs marching in cold-weather gear.[4] By the end of that summer, the base was operational and ready to house thousands of soldiers, a wing of strategic bombers, assorted fighter jets, massive radar stations, and eventually, nuclear weapons, along with a few scientists studying the ice sheet and the frozen ground.

On August 8, 1951, a plane carrying General Lewis Pick, Chief of Engineers, Army Corps, landed on Thule's newly completed 7,000-foot-long runway.[5] A week later, the sun would set for the first time since late April. Pick's men, along with thousands of civilian contractors, had built Thule Air Base in the short Arctic summer—an unprecedented achievement in the face of sea ice that refused to melt and permafrost that did. Pick had told American business leaders, "If that [the building of Thule Air Base] can be done, the construction industry and the Corps of Engineers will have broken the ice waste of the far north."[6] Pick was in Thule to inspect and celebrate that Arctic breaking.

As construction workers returned home to the States in the fall of 1951, the story began to leak out in newspapers around the country, including the *New York Times*, which proclaimed that men were returning home from an unspecified Arctic location to "thaw out their frozen assets." Thule went fully public a year later—first on paper and then in movies.

In September 1952, an amphibious military landing ship dominated the cover of *Life* magazine. There, a silhouetted man waving two flags guides the ship to shore beneath a headline that reads "The Biggest Secret Operation since D-day." Just below, a map shows the distance from Thule to every major Russian city—2,780 miles to Moscow; 1,860 miles to Murmansk, the Soviet submarine base on the Barents Sea. The *Life* article documents the unique challenges of Arctic construction: building a runway on frozen ground, welding massive aviation and diesel fuel storage tanks, sinking piers for foundations to keep warm buildings from melting the frozen soil below. At 75.5 degrees north latitude, the ground is frozen solid for all but a short summer, during which the upper foot or two may soften into impassable mud. Such permafrost and its active layer (the mud) was everywhere in Thule and featured prominently in every construction decision.

But by the last few pages of the piece, Thule almost becomes just another place to live for the Americans and few Danes stationed there. There were religious services for Catholics, Protestants, and Jews, which *Life* reported as "well attended." Baseball was a "major amusement," featuring teams such as the Frigid Digits, the Wet Sox, and the Little Siberians. There was a local radio station and a gym. Thule's television station had the call letters KOLD, and there was skiing in the summer on the margin of the ice sheet. The pay was excellent. Many men made more than $1,500.00 each month, and *Life* reported that "even the janitors draw $226.95 a week." The enlisted men's bar sold only beer, and only one bottle at a time. But it was open in the morning from 6:00 a.m. to 10:00 a.m. for those just getting off the night shift. With the exception of two women, who were the wives of the Danish and American chiefs of the weather station, living at Thule was for men only—except for visiting entertainers and pinups.

The Inuit experience was much different. In 1953, the U.S. military forcibly removed the native Greenlanders, whose traditional land included Thule, to Qaanaaq, eighty miles away—ostensibly to protect them from American influences and disease. Their new site was empty, and the houses provided by the military, built only for the people who agreed to leave without protest, were inadequate for the harsh winters.[7] Many Inuit spent that first winter in tents. Some didn't survive the cold. In the 1990s, 187 people exiled from Thule, along with 500 of their descendants, brought suit in a Danish court asking for $38 million USD in compensation, the removal of the Thule Air Base, and the return of their traditional lands. The court granted each of the Inuit the equivalent of $2,750 and for the group, another $80,000.[8] The Inuit regained access to a small percentage of their homeland, but Thule Air Base remained. In 2023, the American military renamed Thule as Pituffik Space Base. The claim was that the renaming reflected the "rich cultural heritage of Greenland and its people and how important they are to the sustainment of this installation against the harsh environment north of the Arctic Circle."[9]

The same year the military removed the Inuit from Pituffik, the U.S. Army released *Operation Blue Jay*, a made-for-TV movie.[10] In full military garb, the narrator closes the episode by musing patriotically on Cold War tensions. He looks intently at the TV audience as he speaks.

> We cannot read the minds of those dark and shadowy figures who brood on war and conquest but perhaps the thought of this colossal air base has caused them to falter in their plans for aggression. Perhaps, they have read a different kind of warning in the miracle of Blue Jay, the warning that there is nothing, absolutely nothing, impossible to American resourcefulness, hard work, and plain guts. That there is no problem or enemy, nature or man-made, that this country cannot defeat in its stern resolve to protect its freedom.

In 1954, *Operation Blue Jay* received an Oscar nomination in the Documentary Short Subject category. It lost. Walt Disney's *The Alaskan Eskimo* won that year. The frigid Arctic and the Cold War were on America's mind largely because the shortest distance from Moscow to Washington was over the pole and across Greenland. The island was as critical for offense (launching attacks) as it was for defense (alerting the lower forty-eight states to incoming Soviet missiles and bombers). That's why Greenland became so important to the United States military and the American psyche.

The Deadly Mantis is a B-grade 1957 horror movie.[11] The plotline features a football-field-sized praying mantis emerging from the Arctic ice. Soon, the big bug is snacking on American military men in remote camps and flying over the Arctic. The mantis then heads south. Why? Climate change. The movie's token scientist, a paleontologist, suggests that the climate was far warmer in the bug's past life—a million years or more in the past. Now that Earth has cooled, the equator better suits the insect. On its way south, the bug scales the Washington Monument before heading off to New York City, where a military team delivers a deadly chemical agent—a triumph of American superiority.

The movie is a window into the way Cold War America saw the threat of Communist "over the pole" aggression. It showcases both American domination of the hostile Arctic environment and the country's vulnerability. Some speculate the mantis represents the Soviet threat; others suggest that it and other mega-monsters in movies of the time reflect the public fear of radiation-induced genetic changes.[12] Although the emergence of the Deadly Mantis from the ice might have once been a parable for the threat of a Soviet attack across the Arctic Ocean and North Pole, today the mantis better represents a monster of our own making: changing climate and melting ice.

WIND AND BLINDING snowstorms strike Greenland and its ice sheet during every season of the year. In the 1950s and 1960s, the military called these blizzards "phases" and routinely prepared soldiers for conditions that could become life-threatening in minutes. A level-one phase was dangerous. A level-two phase was life threatening, and travel with a buddy was mandatory. A level-three phase was survivable only in a shelter. At Thule, the military developed phase rescue plans. Phase shelters, positioned along the base roads and equipped with food, cots, heaters, and radios, saved lives when the storms hit.

The Thule experience built on what Wegener's men had discovered several decades before and the knowledge the U.S. military had gained in East Greenland during World War II: get people out of the wind and cold and into the ice and snow so that they could live and work successfully on the ice sheet. The Ice Cap Detachment weather station from 1943 was the American prototype. It was a prefabricated building that the military engineers designed to "become completely covered with snow during the first year. . . . Extra sections of smoke pipe and ventilation pipe were provided for use when the house became deeply buried."[13]

In 1948, a French team, led by Paul Victor, launched a multiyear expedition to survey the Greenland Ice Sheet. Victor had joined the U.S. Army Air Corps after the Nazis invaded France and rose through the ranks in Alaska and Europe.[14] In 1949 and 1950, eight members of his Expéditions Polaires Françaises overwintered in a well-insulated, prefabricated cabin buried in the snow.[15] Their camp, which they called Station Centrale, was at the summit of the ice sheet, almost exactly where Eismitte had been nearly two decades before. A generator provided power and illuminated a UV light in lieu of sunshine during the long winter.[16]

Around the cabin were 450 feet of snow tunnels connecting it to science laboratories and storage facilities. As 10 feet of snow accumulated over the winter, the weight of the snow compressed the tunnels beneath it. Their roofs sagged from an original height of 6 feet to 4 feet, and the men resorted to crawling from one area to the other. In the spring, they re-dug the tunnels. When an American team visited Station Centrale five years later, they located the entrance to the camp, but found it crushed and in ruins.[17]

Using more than 600 seismic shots through the ice, Victor's team showed that the center of Greenland was a bowl surrounded by mountains around the margins. The ice sheet was as much as 11,000 feet thick. Beneath it were icebound canyons, some several hundred miles wide and extending up to 1,200 feet below sea level. Using these data, the team calculated the volume of the ice sheet for the first time, showing that it held enough water to raise global sea level about 24 feet—a value that has stood the test of time. The men also used thermal drills to melt their way over 160 feet into the ice. Their mechanical drill reached a depth of nearly 500 feet.

A bucket auger carved a shaft 3 feet wide and 100 feet deep. Victor, lowered into that hole, recalled that "in the light of my electric torch, yearly layers of compact snow glittered like diamonds. As a tree's rings tell its history, so layers of snow and ice reveal the ice-cap's past. Here, before my eyes, was the record of much more than a century of history."[18]

The logistics of supporting Expéditions Polaires Françaises built on American operations in World War II. Supplies were delivered by planes from Iceland, flying only twenty to thirty feet above the snow. Each airdrop delivered five tons of supplies directly to the ice; the pilots reserved parachutes for delicate scientific instruments. After the last airdrop of 1949, the men extracted jerricans—containers typically filled with gasoline—from the snow. But one was different: it was "painted red and labeled 'Pinard.' *Pinard* is a French slang word for red table wine." Victor later learned it was a gift from the French Antarctic expedition. One day in July 1950, all forty-seven members of the Greenland expedition gathered at Station Centrale. "Never had the central ice camp seen so many living creatures in its hundreds of thousands of years." The undersnow cabin had only eight beds.[19]

The next "in-ice" camps, built by the Americans, were an order of magnitude larger and were built differently. They were "designed for under-ice living" in "the soft snow of the Greenland Ice Cap."[20] During the summer of 1953, engineers and ironworkers installed a pair of the prefabricated camps, which became known as Site 1 and Site 2, 260 miles apart in the northwestern corner of the Greenland Ice Sheet. Site 1 was 120 miles north of Thule at an elevation just shy of 4,000 feet; it was only 9 miles from the ice margin. Site 2 was farther inland and higher, 230 miles east of Thule at an elevation of almost 7,000 feet. The bases were a joint venture between the U.S. Air Force, which staffed them, and the Army Corps of Engineers, which did the engineering.[21]

The camps' outer skins were rigid corrugated steel tubes, which, when bolted together, were seventy-five feet long and eighteen feet in diameter. Three tubes of this size, connected by narrower tubes just tall enough for men to walk through, made up each camp. The Army designed the tubes to mimic the pressure hull of a well-ballasted submarine, although reporters and the occupants offhandedly referred to the tubes as "sewer pipes."[22] The goals of their design were to minimize deformation from the crushing weight of snow that would accumulate above each tube, and to encourage slow, even sinking of the camps into the firn below. Engineers installed pressure cells and measured snow-induced stresses all around the tubes. They marked survey points on the tube walls and measured each point repeatedly, graphing the data. In the first year, the tubes deformed less than an inch. They were working as designed, pushing back against ever-increasing snow loads.

Each site housed sixteen men. Insulated buildings inside the tubes included barracks, a mess hall, a kitchen, and a recreation room (which offered Ping-Pong and movies). There were washing machines and homemade barber chairs. Diesel generators provided power and, for ingenious (and bored) airmen, a source of warmth to ferment wine made from rehydrated raisins.[23] The temperature inside the tubes was a balmy 72°F. Outside, it could be −50°F. An Air Force newsreel, released in 1956 after the bases were no longer top secret and one had closed, explained how "a hearty group of airmen beat the weather by living literally under it in a unique housing facility."[24]

On-ice construction didn't come cheaply. The sites each required 2,200 tons of cargo to build and equip, much of it airdropped from bulky C-119s—an early cargo plane, aptly nicknamed the Flying Boxcar. Supplies and people came later on ski-equipped C-47s, military versions of the civilian DC-3 that popularized commercial air travel. It took ninety men one full summer to build each camp. The military delivered all supplies by air because there were no reliable over-ice means of transport to the remote camps. The C-119s would fly low over the ice—perhaps fifty feet off the surface—and drop material, including barrels of oil, onto the snow.[25] Not only was this method energy intensive and expensive, but the cargo sometimes didn't survive impact intact. All in all, each camp cost the

A C-119 Flying Boxcar airdrops fuel and supplies while flying only a few tens of feet above the snow during the 1950s. *U.S. Army photograph.*

U.S. military $1.6 million to build in 1953 (about $17.5 million in 2024 dollars).[26]

The camps at Sites 1 and 2 were identical, and they were both in the ice sheet's accumulation zone. At installation, the camps stood mostly above the ice sheet, with only their lower sections buried in the past season's snow. Soldiers entered and exited through doors on the sides of the tubes. Two years later, snow had completely buried the tubes and the doors. From then on, the men went in and out of hatches at the tops of towers that had once protruded awkwardly above the tubes. For the next several years, the engineers added extensions to the towers, raising them to keep up with the ever-deepening snowpack.

Waste disposal was a challenge. The Army did on the ice what most people in the 1950s did on land. Solid waste went to the dump—at Sites 1 and 2, the dumps were trenches excavated in the ice sheet, which falling and blowing snow soon covered.[27] Mimicking terrestrial septic systems, wastewater flowed into holding tanks under the latrine and the mess hall. From there, pumps moved the sewage through a pipe into a pit about 100 feet away from the tubes. Originating from the heated camp, the wastewater was warm—the Army estimated 50°F–55°F.[28]

As soon as the pipe discharged the sewage into the snow, the warm water began melting its way downward through the firn. Within five days, the sewage pit was 53 feet deep. A few months later, the pit extended 83 feet below the snow surface. The Army sent in engineers to investigate with tape measures, lanterns, and a bosun's chair in which one man, suspended on a rope, dangled in the pit, 23 feet below the snow surface. The sides of the pit at the warmer Site 1 were irregular. There, thick ice layers, the result of summer melting near the coast, resisted the warm sewage. The pit walls were smooth at Site 2, where summers were much colder. The snow and firn at these two sites recorded a different history of summer melting and percolation of surface meltwater into the snowpack and firn. Ice cores would later reveal similar ice layers indicative of warm summers in Greenland's past, but without the scent of sewage.[29]

Sewage problems were evident early on. At Site 2, one of the wooden, stave-built sewage pipes sprung a leak. The escaping wastewater melted the snow supporting one of the tubes, and it began to list. A quick repair to the pipe stabilized the tube. For the next several years, during which Sites 1 and 2 remained active, about 1,500 gallons of wastewater poured into the ice every day.[30] A quick calculation suggests that the U.S. military left behind at least 1.7 million gallons of now-frozen sewage and graywater at each site.

The sewage flowed through the permeable snow and then, as it cooled, froze. Later drilling campaigns brought up cores described as "soft core, clean, no odor" and "hard core, slight color, foul odor." When the cores melted, technicians measured conductivity (resistance to passing electricity) in the resulting water samples. The physical, chemical, and olfactory descriptions matched well. Water from the foul-smelling, discolored cores had low electrical resistance and thus high conductivity—a giveaway that sewage, high in salt from the men's urine, had contaminated the snow. Soft, clean snow cores with no odor had resistance values a thousand times higher, showing that they contained very little salt.[31] This was a critical finding because it meant that when Greenland's snow metamorphosed to glacial ice, its purity was ideal for detecting elements arriving from afar.

Living on the ice took its toll on the men. In a series of reports, Lydus H. Buss, an Air Force historian, found that stationing soldiers away from family and friends at Sites 1 and 2 led to odd, antisocial behavior and depression. Buss concluded that

> constant darkness for long periods and constant light for long periods both act as depressants. There is a period of expectancy for a change which wears thin quickly when change does not occur. This disturbs the normal sleep, hunger, and other bodily habits that have been ingrained for years. It occasions a feeling of confusion that transmits itself in a desire not to get up in the morning, the appearance of hunger at unusual hours, and actual changes in bowel habits. The combined effect is one of lassitude. During the relatively short periods when light begins to appear, we find people excited, and frequently running outdoors just to look. When light appears constantly, boredom soon sets in again and the opposite occurs, confusing all the previously established habits.[32]

A flight surgeon called the constellation of behaviors the Thule effect, describing it as "ennui occasioned by a combination of weather, darkness, and lack of diversions of a type to be found in a civilized community." Perhaps it was the reason why the men stationed at Sites 1 and 2 repeatedly asked for more Ping-Pong balls and movies.

Food under the ice wasn't pleasant either. The cakes were flat. The bread wouldn't rise. The cooks had little experience. Worst was the water. As a recently declassified report from the time suggests,

> One reason that the fruit juices were so well received was the unpleasant taste and smell of the water. Water for these sites was obtained by melting snow in a large, fuel-powered melter. Over a period of months the water began to assume the taste and smell of Diesel fuel. Fumes from the constantly running Diesel engines at each site blew over the snow in the area of the site and permeated the snow. The snow melter itself imparted

Inside the unheated but well-lit metal tunnels of either the Site 1 or the Site 2 sub-ice camp, airmen move supplies in the early 1950s—most likely 1953, just after construction, because there is no frost in the tunnels. *U.S. Army photograph.*

a slight diesel flavor to the water, and after a time the water was nearly undrinkable.[33]

The Air Force's response in 1954 was dozens of recipes tested at high elevation.[34] The Army's response half a decade later was to use nuclear power in place of diesel. In between was Project Jello, a 1955, 100-day expedition across Greenland where men did glaciological research while testing new Arctic foodstuffs—most of which were frozen or dehydrated and all of which were airdropped. The men most liked potato granules. The cabbage was controversial.[35]

Like Thule Air Base, Sites 1 and 2 remained top secret for over a year before the military introduced them to the American public.[36] In December 1954, by then year-old photographs of the tubes standing proud on the ice sheet graced the pages of American newspapers.[37] The articles had titles like "Army Gets Mole Type of Arctic Tent," "New 'Snow Mole' Units to House GIs in Arctic," and "Icy Quarters Built to Sink: Air Force Discloses Unique 'Homes' in Greenland." In February 1955, *Real*, the "Exciting Magazine FOR MEN," published "Our Secret Base under the Polar Ice," which opened with a two-page cutaway sketch of the tunnels under the ice filled with men and supplies, a C-47 plane under a starry sky, and men in the snow unloading a Weasel and passing the cargo down a hatch to the base below[38]

Real's story of Greenland's secret "Base X" acknowledges that "to build a dwelling there means you must lay down the strongest possible foundations. But you can't put a foundation in the Ice Cap. What appeared on the surface to be an unmoving mass actually is in constant churning ferment. Thousands of cubic tons of ice grind and gyrate in contortions that would chew any foundation to bits." Over a late-night coffee at Thule with Pick and Morton Solomon, who led the Army engineers in the Arctic, Victor had offered the solution: "The trouble with you Americans is that you think in terms of battleships when you should be thinking of submarines." This remark kept Solomon up all night thinking, "If boats go under water, why not buildings under ice?"[39] Thus Sites 1 and 2 had a French origin story—which was fitting, because Victor had built Station Centrale under the snow.

The military pitched the undersnow camps as Arctic weather stations. That wasn't the whole truth. Recently declassified documents revealed that Sites 1 and 2 were radar installations that happened to collect weather data.[40] The camps were "gap fillers" positioned to fill the holes in American radar coverage between larger, more powerful, stations.[41] If you look closely at the 1956 Air Force newsreel, the fabric-covered radome, which protected the radar antenna from the elements, is clearly visible.

Ultimately, the undersnow structures at both Site 1 and Site 2 saw only a few years of service. The military abandoned Site 1 in 1956. They abandoned Site 2 on July 1, 1957. There is no record that the Army or the Air Force removed any of the infrastructure, or the sewage. Even without maintenance, the rigid metal tubes resisted the pressure of the snow on and around them. In 1965, twelve years after their installation, the last team of surveyors visited the tubes at Site 2. By then, more than forty feet of snow covered them, yet deformation was minimal.[42] Buried more than a decade, they'd become slight ovals—less than a foot shorter and less than two feet wider. The Army had designed a sub-ice structure that worked, but it was costly and energy intensive to install and maintain because all the supplies had to be airlifted across the ice sheet. The military went looking for an easier, cheaper way to build and supply ice-sheet bases—ones that were large enough to hold hundreds of men.

For millennia, Greenlanders have built shelters from local materials—snow for igloos, and sod for homes where there was no ice. After the Army's experience with Sites 1 and 2, its engineers began to do the same, but at a much larger scale. To do so, they needed to know how the ice and the permafrost underlying it behaved—specifically, how these frozen materials deformed under pressure and over time.

The first experiments occurred at Tuto, a satellite camp on the outskirts of Thule. The name, which stood for "Thule Take Off," referred to the ramp that the Army built there, which allowed wheeled and tracked vehicles to climb up the steep marginal zone and onto the ice sheet. At Camp Tuto, engineers began tunneling into both the solid glacial ice and the frozen earth that lay beneath the ice sheet. The military had two goals: they wanted camps that could survive a nuclear attack, and they wanted to build a rail line in or under the ice sheet to speed the transport of people, machines, and munitions across Greenland.

Bader and the Army both had a keen interest in ice tunnels and how they closed over time.[43] The Army Corps of Engineers started digging a tunnel into the ice at Tuto in 1955, at first by hand with pickaxes and with only Coleman lanterns for light. They soon added explosives to speed the work. But using explosives had a downside: the ice fractured when hit by the shock waves from the explosions (which acted like a hammer strike), and some of the ice walls collapsed after the blasting was complete. But the tunnel remained useful, and in two years it extended more than 600 feet into the ice. Near the tunnel's end, the engineers dug a large chamber 60 feet across and more than 25 feet high—large enough to hold "military equipment."[44] The Army dug a pair of ice pits at the side of the tunnel and stored 110 gallons of aviation gasoline in each. At the time, the project was merely proof of concept, but there were grandiose visions for the future. Summarizing their success, the Army engineers wrote,

> It appears quite feasible that under-the-ice installations could be used for storage, guided missile launching bases, fuel dumps, artificial ramps for access to the ice cap, disposal for nuclear waste products, access through heavily crevassed areas on the ice cap, air raid shelters, and even entire military bases. It is felt that an entire air base, with the exception of runways, could be built under the ice.[45]

The May 11, 1956, issue of *Collier's* magazine took the Greenland tunnel story public with an article titled "Subways under the Ice Cap."[46] The six-page spread, which detailed the how and why of such an engineering wonder, reached more than 2 million people. The map on the opening page depicts a 600-mile-long rail line slicing diagonally through the far northern reaches of Greenland's ice sheet. A cross section below shows a schematic track under the ice. The ice tunnel featured by *Collier's* was just wide enough for a person to walk through, but the goal was ambitious: to extend the military supply line to Peary Land, an ice-free zone along the coast of far northeastern Greenland. In the 1950s, the sea ice there never melted, so building and supplying a base like Thule (which required an armada of cargo ships) was a near impossibility—unless there were a train to bring in people and materials.

Northeastern Greenland was hundreds of miles closer to the Soviet Union than Thule and was thus a strategic choice for the next large American base. "Perhaps even more important," *Collier's* proclaimed, "missile launching sites could be built only a few hundred miles from the pole, sites from which even an intermediate range missile could reach Russian bases in the northland from the Finnish border to the East Siberian Sea." This was not a top secret plan like the building of Thule. It was there for the American public, and the Soviets, to read on the pages of a widely circulated weekly magazine.

In 1958, the Army replaced the pickaxes and explosives used previously for ice excavation at Camp Tuto with an electrically driven machine designed for coal mining. It was much faster and much gentler on the ice. By the time the engineers finished, a new tunnel penetrated more than a thousand feet into the ice sheet. It was about twenty-five feet wide and could easily accommodate vehicles.[47] The Army drew up plans to install a twenty-five-man camp in the new tunnel and cut broad galleries at its far end for the buildings.[48] The tunnel camp would have diesel generators, ventilation, heated buildings, and a mess hall, and would provide blast and radiation protection should a shooting war start.

The ice tunnel was a natural laboratory in which Army engineers and civilian scientists learned how ice sheets work.[49] The engineers wanted to know practical things, such as how quickly the walls closed in. They learned that ice deforms rapidly, on the order of a foot or more per year.[50] The scientists brought in specialized equipment to measure the crushing, flexural, and tensional strength of ice, as well as more esoteric material properties like its dynamic modulus of elasticity, all of which are useful for predicting how ice will deform under pressure and vibration.[51] Today, these data underpin models of ice flow on much larger scales, including the entire Greenland Ice Sheet. Such data are critical to siting and interpreting ice cores, which require flow models of the ice sheet to help date and constrain the source of the ice.

After completing the science, the Army built a camp in the ice tunnel. In October 1961, American and Canadian military brass and Arctic explorers were touring the nearly finished tunnel camp when a major early-season phase struck their bus just as they left the tunnel. The blizzard was so severe that they couldn't travel at all, so the visitors crawled back to the ice tunnel through snow driven by hurricane-force winds. They thus became the first people to spend a night inside the ice tunnel.[52] A few months later, two dozen soldiers occupied the camp for three months.[53] By May 1963, when Americans glimpsed the tunnel's scalloped walls in a *Life* magazine

The ice chapel, where chaplains held religious services, was a famous and well-used site in the ice tunnel at Camp Tuto. *U.S. Army photograph taken by SP4 Jon Fresch, June 12, 1965.*

spread, the Army had abandoned the tunnel camp, but its repeatedly photographed ice chapel and altar remained.

Tunneling through the frozen glacial till under the ice sheet presented a different engineering challenge. The weight of the ice above compresses and compacts the till, leaving a strong, resilient, and difficult-to-excavate material. Subglacial water often saturates till. And at Tuto, low temperatures froze the mixture so that it behaved more like rock than like ice or sediment. The Army engineers gave frozen till the nickname "permacrete"—an apt mashup of *permafrost* and *concrete*. The material was so strong that the engineers made their own by mixing water and sediment. They used the material, which they claimed had four times the strength of concrete and set faster, to make fuel tanks, structural beams, and a table and chair.[54]

The portal at the entrance to the permafrost tunnel read "Tuto Salt Mine, Little Siberia, Greenland." Inside, percussion drills pounded three dozen holes into the frozen till face at the end of the tunnel in a time-tested pattern used for mining rock. The permafrost miners then loaded these holes with charges, and the explosion knocked off a foot or two of material, which they mucked out in mining cars that ran on narrow-gauge rails, haulage equipment left over from the initial ice-tunneling operation a few years before. Mining continued for weeks. When the summer season ended in August 1959, the men had driven a tunnel 9 feet tall, 10 feet wide, and 300 feet long into a frozen hillslope a mile from Camp Tuto.[55]

It wasn't a cheap endeavor at almost $200 per foot in 1959—the equivalent of $2,000 per foot today. Using 1959 technology, tunneling 600 miles under the ice sheet to northeastern Greenland would today cost $6 billion and take more than a thousand years. Permafrost tunneling was clearly not the optimal way to build bases or run rail lines under the ice sheet.

In late June 1954, Bader and his SIPRE team arrived at Site 2. Snow hadn't yet fully buried the metal tubes housing the weatherman and radar operators. The team moved into Camp Fistclench, a collection of insulated fabric buildings, called Jamesways, that looked like olive-green Quonset huts sitting on the snow surface. They carried plans for an undersnow arrangement of tunnels, trenches, and a shaft that dwarfed Eismitte.[56] The ceilings were 14 feet high. The shaft was 100 feet deep with another 58 feet of depth added by a hand-drilled hole. A 60-foot-long tunnel sloped downward until it was 20 feet below the snow surface.

SIPRE came to Greenland with snow shovels, snow augurs, and an assortment of electronic equipment for measuring snow temperature and deformation. The Army provided a half-dozen men to help. They had only two power tools—both chain saws. One ran on gas, the other on compressed air. Men dug and hauled the snow by hand. Their creation earned the nickname "Rabbit Warren."

By August, the men had finished excavating and instrumenting the undersnow maze. Thermistors measured temperatures in the shaft and in the hole bored beneath. Load sensors quantified the stress on tunnel walls. Strain gauges embedded in roof arches tracked deformation as drifting snow buried the trenches. In 1954, the men collected data by hand; there were no automated data loggers. At the end of the summer, all but one of the SIPRE men flew out of the camp, leaving Gunther Frankenstein of the First Arctic Engineer Task Force behind. He spent the winter of 1954–1955 living in the ice (alone, it seems), recording data from many dozens of sensors. Frankenstein's work was critical to the Army's goal that in "five years," it would have "satisfactory design criteria" to build "all kinds of surface and subsurface military installations on high polar ice caps."[57] The Army kept measuring snow and ice deformation in the snow tunnels until 1965.[58] All of these data helped Bader, and the engineers working with him, to understand the mechanics of snow and firn as conditions changed over time.

In the meantime, the military expanded Site 2 during the summer of 1955. It built a snow runway; now at least some visitors could comfortably get to camp in an hour rather than enduring a rough several-day ride over snow from Thule. Along with the runway, the Army brought in a specialized snowplow from Switzerland: the

A Peter Snow Miller blowing snow over the different types of roofing that the Army tested at Camp Fistclench, near Site 2, in 1955. *U.S. Army photograph.*

Peter Snow Miller. Such rotary plows cleared deep snow from passes in the Alps not far from where Bader grew up. Indeed, he introduced the Snow Miller to Greenland.[59]

That summer the Army used the Peter Snow Miller to cut 500 feet of trenches into the snow.[60] After the straight-walled trenches were complete, men installed and tested a variety of different roof structures. Once the forms were in place, the blew milled snow over them. Within days, the "Peter snow" hardened sufficiently, and in some trenches crews removed the underlying forms. Left behind was a snow roof that grew stronger over the next couple of weeks.[61] The milled snow hardened quickly because it had grains of many sizes and because it was much denser than natural snow. Under stress, Peter snow resisted deformation several times better than natural snow.[62] The Snow Miller made in-snow construction at the scale of military bases a reality. The snow it milled became a strong, indispensable, and most importantly, native construction material for use on the ice sheet—no transportation required.

In 1957, Peter Snow Millers "cut and covered" snow trenches to create the first military camp inside an ice sheet.[63] The Army called this new construction, located near Camp Fistclench, the "undersnow camp." Roofs were mostly timber (although a few were metal, and a couple were hardened snow arches). The Army erected insulated Jamesways inside as barracks. A few dozen soldiers spent the summer living in the ice sheet (several times that number lived in the surface camp).[64] Under the ice were a mess hall, kitchen, latrine, rec hall, sleeping quarters, and a dispensary with on-site medical personnel. Diesel generators provided power, and diesel snow melters supplied water, which tasted like fuel, just as it had in the tubes at Sites 1 and 2. The Army used the undersnow camp for several years, abandoning it in 1960. When engineers went back in 1963, the timber roofs had collapsed, the trench floors had heaved, and snow and ice deformation had narrowed the tunnels. The metal tubes of Sites 1 and 2 fared far better. Metal resists deformation better than snow and ice.

Thule, Tuto, Site 1, Site 2, and Camp Fistclench were more than military bases. They had become home to scientists studying all things frozen. Much of what they did was "applied science" because it related to the behavior of snow, ice, and permafrost, which was relevant, at the time, to the military's mission. But the influence of military science in the Arctic far outlasted any of the polar camps. Today, work done with and for the military in the 1950s and 1960s remains the foundational knowledge on which our modern-day understanding of ice and snow science relies.

In the background, but critical to communicating polar science inside and outside the military, was a women, Lucybelle Bledsoe, a technical editor with a graduate degree in English whom Bader hired in 1954. For the next twelve years, she edited just about every report and publication about glaciology that came out of SIPRE and CRREL, the Cold Regions Research and Engineering Laboratory, which subsumed SIPRE in 1960. As James Bender, a SIPRE scientist, wrote of Bledsoe,

> Her editorial work extended far beyond grammatical corrections and adherence to style. She could spot incorrect equations, slipshod terms and desultory sentences. One of her unique talents was her ability to recognize faulty logic even on very technical matters. Many researchers (both young and old) have seen their raw, sometimes confused manuscripts transformed into beautifully simple, well presented reports.[65]

Bledsoe was a lesbian working for the Army Corps of Engineers in the 1950s and 1960s when homosexuality was at best a risk in government service.[66] She was not alone at SIPRE, however. Bader hired several administrative staff who were also in same-sex relationships, although none of them apparently in the open.[67] Not a single report or journal article includes their names in the author lists, or even in the acknowledgements.

In 1956, six years after coring the Taku Glacier, Bader drilled into the ice again—this time at Site 2 in northwestern Greenland. Site 2 was most things the Taku glacier wasn't. It was at high elevation. The climate was far colder. The snow at Site 2 rarely melted, and rain was a rarity, which mattered because melting blurred the record of climate history preserved in ice. Instead of tents, the men now slept in insulated Jamesways. Heat came from kerosene-burning potbelly stoves. There was a mess hall at the camp and flights back and forth to Thule. The deep, lasting chill, and the extensive logistical support of the military, drew Bader, his colleague at SIPRE, James Bender, and twenty-seven-year-old geologist Chester Langway to Greenland that summer.

After a six-year stint in the Army, Langway had enrolled at Boston University in 1952, supported by the GI Bill.[68] In 1956, the Army Corps of Engineers hired him fresh out of school with his master's degree in geology. Langway's entry into Greenland and ice-core science seems rooted in pragmatism. In a 1989 interview, he remembered that "they told me that if I went to Greenland, where they were building the program up, I'd get double pay, because I'd work 78 hours a week. And since I was a student, I was broke, so I said, Fine."[69]

When the drilling rig and drillers arrived at Site 2, the Army used a Peter Snow Miller to excavate several trenches. The drilling rig went into one. The other was a science trench, similar to the one Bader and Wasserburg used on the Taku Glacier, with one exception: at Site 2, the temperature in the trench was a cold and steady 10°F. The men erected a thirty-foot-high drill tower and covered both trenches. The top of the drill tower poked through the trench roof. A heated Quonset hut, designated as a laboratory, held equipment for analyzing the core when it was time to melt the ice samples.

In 1956, the scientists used off-the-shelf Army drilling equipment. They equipped the rig with compressed air to blow chips, made by the drill as it cut through the ice, out of the borehole. To keep the compressed air from leaking out through the uppermost, younger and more porous, snow and firn, the drillers drove a metal casing down about 150 feet. Below that depth, the snow had turned to solid ice, effectively containing the air. As the borehole got deeper, removing the chips grew increasingly difficult, and drilling slowed. Thus the Army learned that compressed air was not an optimal fluid for drilling deep ice cores. They had no way to know this ahead of time, since no one had tried to drill an ice core this deep at a site this cold.

Unlike the Taku drilling rig, which used a wireline to retrieve cores, the Army's Site 2 rig used a solid metal drill stem, made up of a string of threaded, rigid metal rods, each twenty feet long. This made for slow drilling because every time the core head filled with a section of ice core, the men had to winch up the entire drill stem—laboriously unscrewing each of the drill rods, one by one, as they came out of the hole. They then emptied the core head of ice (much of it shattered or missing) and lowered the drill stem back into the hole, one rod at a time.[70]

Tripping in and out of the hole in the ice could take hours. The deeper the team drilled, the longer it took. By the end of the summer, even working double shifts, they could retrieve only two ten-foot lengths of core per day. The men drilled 965 feet into the ice at Site 2, at which point the drill bit got stuck and they could not get it out of the hole. At the time, theirs was the deepest ice core ever drilled.

The next summer, Langway, Bader, Bender, and the drilling team came back to Site 2. Failing to retrieve the bit, they moved the drilling rig 35 feet down the trench and started over. In 1957, their hole reached 1,333 feet below the ice-sheet surface. Again, the ice-core quality was variable. Still, drilling to this depth was a stunning achievement, and there was plenty of ice with which the scientists could work. It helped that this new core was four inches wide, twice the width of the Taku core. But the ice under Site 2 was almost 7,000 feet thick, so even their summer of drilling had barely scratched the surface.

After the drillers removed each core segment from the core head, the men moved the long, narrow cylinders of ice to the science trench, where the cores remained safely frozen. First, they cut each core into five-foot segments. Then, one by one, Langway and his colleagues used several pieces of equipment to characterize the cores. They began by laying the cores on a light table. Backlighting revealed layers in the ice. The men described what they saw, recording the details by depth in a core log. Darker layers might contain bits of volcanic ash from eruptions (most likely from Iceland). Slight differences in light transmission, reflecting winter and summer snow, revealed annual layers. Counting those couplets gave an estimate of the age of the ice.

The scientists used a powered band saw to cut the core segments.

First, they split each segment the long way, setting an archive half aside for the future. The remaining half-round they sliced like a salami. They weighed each slice, then immersed it in a displacement bath to measure its volume. Together, those data give ice density—mass divided by volume.[71] Other slices allowed for different analyses, including "bubble pressure," the amount of ancient air released from frozen bubbles when a slice of core melted. Once the men completed core logging and processing, they placed each core segment in a plastic bag, sealed it, and loaded the bag into a shiny, aluminized cardboard tube for storage and shipping. At the end of the drilling season, the 1957 Site 2 core was returned by plane to the United States, kept frozen by dry ice—solid carbon dioxide.

Within a year, Langway published a short paper about the Site 2 ice core.[72] The Herald Tribune News Service picked up the story with a headline that read, "1100 AD Icicle Studied by the Army."[73] Bader's quotes in the news service story show a vision for ice cores that in retrospect seems prophetic.

> Every snowfall and everything that fell with it are, so to say, separately and safely filed for future reference. Natural objects which fell with the snow—such as volcanic ash—are preserved year by year. Scientists who have been monitoring radioactive fallout can go back to the ice caps to measure some things they missed at the beginning. It should be possible to follow quantitatively and qualitatively the degree of atmospheric contamination by industrial activity through analysis of snows down to layers which fell in pre-industrial times.[74]

In analyzing the Site 2 ice core, Langway built upon the intellectual foundation Bader had laid. It was Bader who first imagined how to tease climate signals out of long cores of ice, seeking them in dust, oxygen isotopes, dissolved elements like sea salt, and detailed visual and microscopic analysis.[75] Together with Wasserburg, he vetted those ideas on the wet Alaskan Taku glacier in 1950. Along with rapidly advancing drilling and analytical technology, Bader's inspiration primed the field of ice-core science to take off. In the following decades, it was Langway who would team with others around the world to make Bader's vision a reality.

Langway's paper had only one graph and was only a page long, but near the end is an invitation that proved pivotal: "Interested persons are cordially invited to submit proposals for further studies." Sam Epstein, a young geochemistry professor at Caltech, took Langway up on his offer and asked for more than 400 samples. Langway sent ice cut from four different core depths.[76] Epstein was one of the world's few experts in oxygen isotopes at the time, having developed and refined methods for their measurement while in training, along with Wasserburg, at the University of Chicago.

Geochemists like Epstein used mass spectrometers to measure the mass of both the hydrogen and oxygen atoms that make up each water molecule. In the 1950s and early 1960s, measurement was a slow, manual, and laborious process of moving gas into and out of custom-made instruments. Data were not easy to come by, and only a few laboratories in the world could make the measurements. Epstein ran one of those labs.

To make sense of the isotope data, you needed to know that every oxygen atom has eight protons in its nucleus, and that most oxygen atoms also have eight neutrons (and are thus designated ^{16}O—the number indicates the total atomic mass). A few have nine neutrons (^{17}O) or ten neutrons (^{18}O), which means an oxygen atom can weigh sixteen, seventeen, or eighteen atomic mass units. Put a heavy oxygen atom in a water molecule, and that molecule weighs just a little bit more than its more common siblings. This makes a measurable difference when it comes to evaporation and condensation. The lighter water molecules evaporate a bit more easily than the heavier ones—a phenomenon called density-dependent fractionation. Even more importantly, the behavior of water molecules and their associated oxygen isotopes is temperature dependent. This means that the ratio of ^{16}O to ^{18}O in water, and thus in snow and ice, varies not only as the climate changes, but also with the seasons.

In 1952, Willi Dansgaard, a Danish physicist with his own mass spectrometer, was interested in water and how it moved through the atmosphere. With the ability to weigh individual isotopes in water molecules, he designed an experiment involving a funnel and a beer bottle set out in his Copenhagen yard. For three days in June 1952, Dansgaard collected rainfall, periodically emptying the bottle, as first a warm front and then a cold front passed over the city. His data showed that the oxygen isotope composition of rainwater changed systematically with temperature.[77] This discovery suggested to him that ice cores (because they preserved ancient precipitation) could become paleo-thermometers—a way to read past temperatures. Dansgaard's backyard experiment was a humble start to a sixty-year career with isotopes, especially those preserved in ice.

Dansgaard's data suggested that winter and summer snow, which fall at quite different temperatures, should have different ratios of ^{16}O to ^{18}O. So if you were prepared to measure a lot of samples, as Epstein did (251 in total), you could, at least in principle, identify annual ice layers in snow and ice cores using the seasonal back-and-forth of oxygen isotopes. Counting these swings as you move down a core would give the age of the ice, much like counting rings on a tree. But this method had its limits: first, as you go farther down the core, the seasonal layers get thinner and thinner as the ice compresses and flows outward, and second, no one had ever shown that this method would actually work.

In 1959, Epstein published his results.[78] At depths of about 100 m and 300 m in the Site 2 core, levels at which "most of the grain characteristics of the ice have been obliterated, the seasonal variations in the ^{16}O to ^{18}O ratios are still preserved. Thus, where other criteria failed, it was possible to use isotope data to determine rates of accumulation for ice at least 800 years old." The implications of Epstein's measurements were huge and long lasting. To this day, the detailed chronology of every ice core depends on precise, high-resolution measurements of stable oxygen isotopes.

* * *

THOSE INVESTIGATING THE science of climate change in 1958 were like early adolescents, curious but confused. People like Maynard Miller wondered why many Alaskan glaciers were retreating, and the military was worried about its Arctic operations should polar ice begin to melt.[79] Some thought that Earth was getting warmer, but there was no agreement on why, how much, or what drove the warming. Bader, addressing Air Force scientists and managers, envisioned understanding climate and how it changed over time as

> an interdisciplinary objective which is of first class geophysical interest, and which could perhaps become of prime economic significance in the not too distant future. So I would like to put up for panel discussion the problem of warming up the Arctic (and Antarctic). . . . First there is the gathering, by literature search and further field work, of the material necessary to clearly demonstrate the phenomenon of warming up—its geographical distribution, annual distribution, magnitude and trend. What are its geophysical and biological consequences? . . . Can we predict quantitatively what will happen if the warming trend continues or if it reverses? Will the polar climatic change influence the climate of temperate regions, and how?[80]

Scientists knew that carbon dioxide in the atmosphere had the potential to warm the planet. The physics had been clear for nearly a century, since Eunice Foote's experimental work in 1856. Her paper describes filling glass jars with different gases and exposing them to the sun. The jar with carbon dioxide warmed the most.[81] After World War II, people began moving carbon from the solid Earth to the atmosphere at an increasing rate.[82] Our burning of fossil fuels—coal, oil, and gas—was Foote's experiment on a global scale.

If the concentration of carbon dioxide in the atmosphere rose, the planet should warm. In 1958, Gilbert Plass, a physicist whose research the Navy supported, calculated that if the concentration of carbon dioxide in the atmosphere doubled, the average temperature of the planet would rise 6.8°F—a value known as the planet's climate sensitivity.[83] After almost seventy years of intensive research and reams of data, the most likely value has dropped only slightly, to about 5°F.[84] But the pivotal question remained, was the concentration of carbon dioxide in the atmosphere actually rising?

Accurately predicting Earth's climate future required understanding what happens to carbon dioxide once it enters the atmosphere from smokestacks and tailpipes. If the oceans efficiently absorbed the gas from the atmosphere, global temperature change would be minimal. If the carbon dioxide remained in the atmosphere, the planet's temperature would rise, perhaps substantially. But, in the 1950s, there were few measurements of carbon dioxide in the atmosphere, and even less agreement on how much of the carbon dioxide emitted by fossil fuel combustion ended up in the oceans and how much remained in the atmosphere.[85]

Chemist Charles Keeling filled the data void. As a postdoc at Caltech, Keeling developed equipment that accurately determined the concentration of carbon dioxide in air. By March 1958, Keeling had moved to the Scripps Institution of Oceanography in San Diego, where he encountered Roger Revelle, who had dedicated much of his life to figuring out how carbon moved into and out of the world's oceans. That year, Keeling began making frequent measurements of atmospheric carbon dioxide concentration on the 14,000-foot summit of Hawaii's Mauna Loa. He intentionally placed his samplers on the mountaintop, far from cars, factories, and the carbon dioxide they emit. Over time, Keeling's measurements would provide a way to answer Revelle's most important questions about the climate-ocean-atmosphere connection.[86]

Using his first two years of data, Keeling published a paper that, for the first time, showed the annual atmospheric ups and downs of carbon dioxide concentration. Levels were lowest in October and highest in May. Keeling inferred correctly that his measurements tracked the net photosynthesis and respiration of Earth's plants—

biased to the Northern Hemisphere, where there is far more land (and far more plants) than south of the equator. All summer, growing plants removed carbon dioxide from the atmosphere. In the fall, they dropped their leaves, and decomposition, which releases carbon dioxide, took over. The conclusions Keeling drew were prescient: in a 1960 paper, he equated rising levels of carbon dioxide in the atmosphere with increased fossil fuel combustion, writing, "The observed rate of increase is nearly that to be expected from the combustion of fossil fuel (1.4 ppm) if no removal from the atmosphere takes place."[87]

Keeling's data, and the curve that now bears his name, clearly show the carbon dioxide concentration in the atmosphere rising steadily and annually. When Keeling started measuring carbon dioxide in 1958, neither he nor Revelle had a way to put those measurements in the context of Earth's history. To do that, they needed a record of carbon dioxide concentrations in the atmosphere before people influenced the carbon cycle. In 1949, as Bader sat with a microscope and a backache examining ice from Alaska's Malaspina Glacier, he saw bubbles in the ice.[88] A year later, Bader found similar bubbles in the Taku Glacier ice core. In 1957, James Bender made the critical intellectual leap after identifying air bubbles in the Site 2 ice core: the bubbles, he hypothesized, were tiny archives of ancient air.

In ice cores, Bender saw the way to determine whether people had altered the chemistry of Earth's atmosphere beyond its normal range of variation. Ice-core analysis had the potential to reveal natural fluctuations in atmospheric carbon dioxide—not on the time scale of Keeling's Hawaiian air samples, which do an excellent job of recording carbon dioxide's atmospheric rise over years and decades—but on the scale of centuries and millennia. If scientists could date glacial ice and measure the concentration of carbon dioxide in the tiny bubbles, ice cores would provide context for Keeling's measurements. Bender speculated that the deepest samples from the Site 2 core might preserve air from the Middle Ages, nearly a thousand years ago, well before the start of the Industrial Revolution.

In 1959, Bender shared his ideas about the Site 2 ice core with Ralph Segman, a reporter from Science Service, a nonprofit news organization dedicated to the public dissemination of science.[89] At that time, no one could measure the concentration of carbon dioxide trapped in ice, and the Site 2 core was too short and too shattered to be useful. More than twenty years passed before technology was up to analyzing the ancient air in those tiny bubbles accurately. Bender was decades ahead of his time, and his insight would turn out to be critical. To him, those long cylinders of ice were the Rosetta Stone by which we could decipher the long history of Earth's atmosphere and, by extension, Earth's climate.

Now, the challenge was to collect a core complete enough and old enough to reveal the magnitude and timing of past changes in Earth's climate system. Bader, Langway, Bender, and the U.S. Army science team needed a continuous ice core that penetrated an entire ice sheet. Building on the foundation of applied polar science they'd created over a decade for the military at Tuto, Sites 1 and 2, and Camp Fistclench, the team was ready to collect just such a core. They only needed to find the right place to drill.

Peter Snow Miller undercuts a trench in Greenland's snow during the construction of Camp Century in 1959. *U.S. Army photograph. Robert W. Gerdel Papers, SPEC.PA.56.0022, Byrd Polar and Climate Research Center Archival Program, Ohio State University, Box_2_Folder_29_012.*

WARMTH UNDER THE SNOW

> He has brought his greatest scientific achievement, power from the atom, to the very top of the world, but can he live here?
>
> —WALTER CRONKITE, *City under Ice*

A striking color painting graces the February 1960 cover of *Popular Science.*[1] Two Army surveyors, well dressed for the cold, stand in a long, narrow trench—cut perhaps thirty feet deep into the snow. The walls are white and blue. Between the men, a giant yellow Peter plow deepens the trench walls, blowing the snow out through long, fragile-looking chutes. In the distance, the northern lights, glowing orange, dance across a dark, almost purple, starlit sky. The headline reads, "U.S. Army Builds City under Ice." That city, the military base called Camp Century, was the culmination of a more than a decade of Cold War engineering on and in Greenland's ice sheet. It was the direct descendant of Sites 1 and 2, Camp Fistclench, and the undersnow camp, designed and built on a foundation of applied science that started in Eismitte.

News of Camp Century went public a year before the Army built the camp.[2] Site visits began in summer 1958. In December of that year, the *Cincinnati Enquirer* ran two stories, "Dig Deep for Gracious Living Beneath Greenland's Ice Cap" and "For Survival, Dig In."

Both profile Robert Phillipe, an Army engineer, and describe his talk to the Society of Military Engineers. The photographs are of Fistclench, since its appearance was the closest to how Camp Century would look. They show deep trenches, the entrance to the camp, and an oil-heated Jamesway beneath the snow. This was not a leak of secret information, but a scoop for the paper. The Army had just released the photos.

The Army built Camp Century in two summers, starting in June 1959, at a cost of $8 million.[3] Building the initial camp in 1959 and 1960 required 6,000 tons of construction material, all dragged across the ice by massive diesel tractors pulling sleds; more material came in later years. The Army used Peter Snow Millers to create Camp Century, as they had to excavate the trenches of the under-snow camp at Fistclench. After the Peter plows finished excavating the snow, the crew covered each open trench with an arched metal roof. Then, a Peter plow buried the metal with blown snow. Once the snow hardened, the crew often removed the metal supports and reused them elsewhere.

When the Army built Camp Century, it already knew a lot about Greenland and its ice sheet. Dozens of military scientists had spent the 1950s gathering information about the effects of stress on ice and snow. What they learned, both in the lab, where large presses squeezed samples, and in the field, where the first ice and permafrost tunnels deformed, was a subtle but sobering warning for Camp Century. Unlike the excavations at Tuto, Century was dug in snow—much softer and more deformable than the hard blue ice or permafrost near the coast. The difference was striking. In the Tuto ice tunnel, the roof lowered about a foot per year. At Camp Century, some snow tunnels would close far more quickly—at up to 4.2 feet per year—a rate reminiscent of Station Centrale.[4] The tunnels, carefully milled by the Peter plows, were destined to rapidly deform.

Camp Century was not alone in northern Greenland. Building, maintaining, and running the camp required a complex supply chain to move people and cargo from the American mainland to the coast of Greenland and then to the ice margin and across the ice sheet. Supplying the camp required a complicated multimodal transportation network. Ships and planes moved people and cargo from ports and airbases in the lower forty-eight states through Canada, and eventually to Greenland, and then, using massive tracked vehicles, across the ice.

Critical to this effort was the sprawling American airbase at Thule. It was the supply depot and way station for Camp Century, an essential conduit through which men, materials, and eventually, ice cores flowed on and off the ice sheet. The base, and its lengthy paved airstrip, allowed scientists and their equipment to come and go on Military Air Transport System (MATS) planes from North America. The deepwater harbor hosted a flotilla of cargo ships every summer when the ice broke.

The name "Century" reflects the camp's initial proposed location, 100 miles from the edge of the ice and about 800 miles as the crow flies from the North Pole. This was not where the camp ended up—the ice sheet sloped too steeply for construction at that distance. The engineers kept moving inland along the Army's snow road over the ice sheet until they found the first level place where the snow stayed cold and dry all summer, 138 miles from the coast. Camp Century plus 38 was a mouthful, so the Century name, however inaccurate, stuck.

The camp sat at an elevation of 6,600 feet, where the air was thin and cold and the snow was dry. Rarely, even in the constant daylight of summer, did temperatures creep toward freezing. The dark winters were frigid; −50°F was not unusual, and that was without the wind chill. Averaged over the year, the air temperature at Camp Century was −11°F. Without wind, bare skin would freeze in a half hour. With a breeze, in ten minutes.

Most soldiers and scientists traveled to Camp Century in heavy swings: slow-moving over-ice convoys that began outside of Camp Tuto where the gravel ramp transitioned to the ice road. Massive

Barracks wanigan on a heavy swing to Camp Century equipped with clotheslines, a punching bag, and bunks stacked three high. Photograph published January 9, 1961. *Milwaukee Journal, © Fred Tonne—USA TODAY NETWORK.*

D-8 tractors with extra wide tracks pulled a string of sleds behind them across the ice sheet. Many of the sleds contained cargo, but some were mobile barracks, another was the mess hall, and yet another was the radio sled.

The trip to Century was a three-day, round-the-clock affair. In good weather, the tractors moved at about the speed that a person could walk. In bad weather, the swing would grind to a halt, sometimes for days. Otherwise, the swing stopped only for tractor refueling as fresh drivers came on duty. VIPs and scientists often took the fast train—Polecats, which were smaller tracked vehicles that moved much more quickly over the ice-sheet surface. Flags lined the route, but if the drivers could not see the flags through the snow, the swing waited. The lucky few, when the weather allowed, flew for an hour from Thule or Tuto.[5]

Those living at Camp Century needed water, and it had to come from the ice. The Army Corps of Engineers spent three years figuring out how to get water from an ice sheet where the temperature rarely rises above freezing and, except in the brief melt season, liquid water is scarce. The shortage of water results from the energy required to change ice into water, a phase transition. Melting enough ice to provide one person's daily drinking water needs requires about as many calories as a large fast-food hamburger contains.[6]

For expeditions carrying their own gear, including fuel, it's clear why water was scarce: the men couldn't carry enough fuel to both melt snow and cook their food. Food came first; as a result, early explorers complained of nearly constant thirst.[7] Fridtjof Nansen's party was the first to cross southern Greenland. In 1888, they completed the 260-mile, forty-one-day journey over the ice on skis and on foot, dragging hundred-pound sleds. The only drinking water they had was from metal flasks, filled with snow and carried next to their bodies. Body heat melted the snow. Slowly.

The Army Corps understood the energy demands of phase-change physics. At Camp Century, the engineers harnessed steam to make the ice-water phase transition. It takes almost seven times more energy to flash water to steam as it does to melt the same weight of ice. That equation also works in reverse. When steam meets cold ice, it condenses and releases energy, the latent heat of condensation. That energy melts ice. This is why steam radiators heat up so quickly: the steam turns to water on the cold iron and transfers heat to the radiator and then to the room and you.

At Camp Century, the engineers first used jets of steam (produced by a steam drill heated by diesel fuel) to melt a borehole 230 feet into the ice below the camp. Then, they used more steam to melt a cave 60 feet deep and 70 feet wide at the bottom of the borehole.[8] Every day, more steam melted more ice. Men used the once-glacial meltwater for cooking, washing, hot showers, and drinking. Army engineers called these ice-sheet melt-outs "Rod wells" in honor of the man who designed them, Raul Rodriguez, a native of Puerto Rico.[9] The idea caught on, and now most Antarctic Ice Sheet camps use Rod wells for their water supply.

Because many feet of snow accumulate at Century each year, but that snow compacts as more snow piles on top, camp occupants drank a mix of melted snow that fell sometime between when General George Washington was president of the fledgling United States (1776) and the giant eruption of Indonesia's Krakatau volcano in 1883. That eruption spewed so much ash and sulfuric acid into the atmosphere that the climate cooled and sunsets were brilliant red for months. Evidence of Krakatau's fury resides in Greenland's ice. You can distinguish the ice from 1883 because it is more acidic, and conducts electricity better, than the ice below or above.

CAMP CENTURY WAS truly a city under the ice. As a military base inside an ice sheet, the camp was a unique outpost of humanity in what many have described as a featureless icescape without end. Over a hundred men lived there isolated for months at a time under the snow, cut off from the world by blizzards, magnetic storms, frigid temperatures, and howling winds. Jon Fresch, an Army photographer who first went to Greenland when he was eighteen years old, described living at Camp Century this way when we talked: "I liken it maybe to being on a submarine because once you went down, for the most part, unless you had a job that took you above ground or onto the surface, you had no reason to be up there because when it stormed it stormed, and you didn't want to get lost, that's for sure."

There were two large entrances to the camp's subsurface tunnels (through which vehicles and men entered), cooling towers that dissipated waste heat from the purpose-built nuclear power plant that powered and heated the camp (more on that later), and a series of escape hatches. Otherwise, blinding white snow stretched uninter-

An Army truck under the snow at Camp Century in the summer of 1963. It's outfitted with metal wheels so it can move along the railroad tracks. *U.S. Army photograph, Robert W. Gerdel Papers, SPEC.PA.56.0022, Byrd Polar and Climate Research Center Archival Program, Ohio State University, Box_6_Folder_23_047.*

rupted toward the horizon. Like Eismitte thirty years before, Camp Century was hardly noticeable at the surface.

Yet beneath the year-round blanket of snow stretched a gridded maze of twenty-eight tunnels, each a few tens of feet deep. The military assigned each a number. Most tunnels contained prefabricated buildings (there were twenty-eight buildings in total) raised off the icy tunnel floors by wooden foundations to keep the snow beneath from melting. Main Street, itself a tunnel in the ice, was more than 1,000 feet long and wide enough to accommodate large over-snow vehicles. Tunnel 6 held the mess hall. Six hundred feet to the northeast was a railroad tunnel with a flatbed rail car, a military truck fitted with metal wheels, and over 1,600 feet of track, shaped like a smoothly arching fishhook.

In the relative calm and warmth of tunnel 12, a drilling rig spun incessantly, slowly deepening a hole in the ice sheet. The air was often thick with the smell of the freeze-proof but toxic drilling fluid that kept the hole open. Men in heavy clothes used winches to gently pull up long, skinny cylinders of clear glacial ice. Every time a core came up, scientists and engineers wiped off the drilling fluid and began the initial core processing, carefully noting the core depth and length and then examining the ice on the light table for what might be annual layering or a dark layer of volcanic ash—the smoking gun of an ancient volcanic eruption.

Snow encased the camp, sheltering its occupants from some of the world's harshest weather. Men lived below the surface in well-insulated barracks with heated floors, bunk beds, and comfortable lounge chairs. The science labs were first rate, and there was a library, a post office, and a hobby shop. A store sold modern cameras and a lot of film. There was a darkroom to process that film, used by Army photographers, like Fresch, who documented the camp. The medical suite included an X-ray machine and equipment for basic procedures.

Men at Century were allocated 50 percent more rations than those in temperate regions because of the caloric needs of working and living in the cold. With some people working around the clock, the mess hall served four meals a day, which often included all-you-can-eat steak. The daughter of one soldier told me the story of the family making a special steak dinner for the serviceman's first night home from Greenland—they were excited, but he was disappointed, having eaten so much steak on the ice.[10] Then, there was the windowless latrine. It had a dozen porcelain toilets (back to back, six to a row) and twice as many sinks. One man stationed at Camp Century told me when we spoke that there were more constipated GIs there than you could ever imagine. Because you sat on a pedestal while other people stared, men would set their alarms for 2:00 or 3:00 a.m. so that they could be alone.

In the absence of television, movies, shipped in from Thule, played most days. One soldier claimed to have watched 179 films in a single tour of duty. The airstrip allowed chaplains from Thule to visit on Sundays when the weather was good, or later in the week if Sunday wasn't a safe day for flying. There was a bar where all drinks cost 25 cents, and a large photograph of a woman in a bikini hung behind the bartender.[11] Dansgaard wrote in his diary that, when a clergyman came to the camp, the men turned the picture around to display a portrait of someone fishing instead.[12]

A different pinup nearly killed Joseph Kumbar, a civilian sanitary expert assigned to Camp Century. He'd volunteered to inspect one of the deep Rod wells melted into the ice sheet using the steam drill. This particular well supplied water to the camp.

> Many officers who visit Century make a point of inspecting the Rodriguez steam drill. Most are undoubtedly interested in its technical ingenuity, but others thoughtfully study the big pin up photo carefully fastened to the top. The drill was pulled up and Joe Kumbar started down into the well in a parachute harness. . . . The winch operator became preoccupied—with the distracting photo, GIs speculate—and dropped Kumbar so far and so fast that he came within three feet of being dashed to death against the frozen bottom. He bounced up and down in his harness like a Yo-yo. . . . The controversial photo still hangs proudly in the pumphouse.[13]

The popularity of pinups at Camp Century reflects the conspicuous and intentional lack of women at the camp, with few exceptions.[14] Almost every popular piece written about the camp and nearby support bases mentions their absence. The first book published about Camp Century relates that a common joke was "to tell new arrivals on the glacier that there's a girl behind every tree in Greenland. In case you've forgotten, there are no trees in Greenland."[15] Two teenage Boy Scouts who spent nearly half a year at Camp Century remembered that "the nearest girls were 180 icebound miles away."[16]

For all the joking, there were downsides. Bader, speaking to the Air Force management and dozens of polar scientists in 1961, was scathing in his rejection of the military's cold-region culture of same-sex isolation and poor living conditions.

> Perhaps the most revolting aspect of arctic research is the prospect of deprivation of feminine company, professional and social. Can we humanize the profession by normalization of working and living conditions, or are we to continue to lose our young men after two or three field seasons under completely outdated expeditionary conditions, or soon after they get married? In the last fifty years we have done little more than to shorten travel time. Instead of living in miserably uncomfortable tents, we now live in miserably uncomfortable barracks. This has to change if we are to get good work done during the winter.[17]

CAMP CENTURY WAS both an isolated and an isolating place. Mail deliveries were at best unpredictable. There were no telephone lines on the ice. Servicemen called home from the camp using ham radio. A soldier "ham" would contact another ham in the mainland United States, who would then patch the radio call into a telephone line. Radio operators traded postcards by mail to confirm that they'd made contact across the airwaves. Postcards from Greenland were particularly valued because there were so few ham operators there. The card from Camp Century, which read "From the Land of the Midnight Sun," had an atom symbol with electrons orbiting the nucleus and the words, "Nuclear Powered."[18]

Camp Century was no secret. The military authorized a half-dozen illustrated magazine pieces and three full-length books featuring the camp.[19] Colonel Gerald Homann, commander of the Polar Research and Development Center, was on the TV show *What's My Line* in 1963, while *Our Weekly Reader* had a story about Camp Century for fourth graders. The radio series *Young Ameri-*

cans in Action featured Leon McKinney, the officer in charge of the Camp Century reactor.

The world, and especially Americans, seemed smitten with the idea of people living in and conquering the ice, in the same way they'd admired Peary's quest for the North Pole more than a half century before. Perhaps this was because the Army designed and executed a purposeful publicity campaign for Camp Century, one suitable for the Cold War and not limited to the United States. In 1962, an article in *Spirou*, titled "Le Bon Home de Neige," included six photographs of the camp and the reactor along with two pages of text.[20] Two months later, a full-color, cutaway diagram of the camp was the featured centerfold in another French magazine, *Pilote*, with the title "Camp Century, Cité Polaire."[21]

U.S. Army newsreels also trumpeted "the City under the Ice," showing full-color footage of the camp's construction, including Peter plows and men installing and starting up the reactor. In one newsreel, Langway, unmistakably tall and dressed in Army greens, finishes dinner and casually lights up a cigarette in the Camp Century mess hall.[22] In another, the camp dog, Mukluk, entertains the men. Weasels drive on Main Street and pull men on skis across the ice. Soldiers relax in the barracks and shovel freshly roasted turkey and homemade biscuits onto plates in the mess hall. Camp life in the newsreels seems remarkably mundane.

The most effective publicity ploy was the simplest. It built on a military connection that has existed for more than a century, ever since Robert Baden-Powell founded Scouting in 1906: take Boy Scouts to the Arctic—one Danish, one American—and let them do science and experience military life. The narrator of an Army newsreel suggests a deeper motivation. "It's no longer easy for young Americans to meet nature in the raw. . . . Boy Scouting is one of the few institutions to balance the rather softening effect of our modern way of life," to help young men develop "the leadership they really need in the somewhat jittery insecure world in which we live."

There was fierce competition to become the one American

Boy Scout who would bunk in Greenland. Twelve regional councils nominated a total of 120 Eagle Scouts. Seven finalists flew to New York and went head-to-head in a final, in-person examination. Kent Goering of Neodesha, Kansas (population 3,594 in 1960), was the winner. He spent the winter of 1960–1961, from October until April, a mostly dark and always cold time, in Greenland alongside Soren Gregersen, a Danish Scout. Both were seventeen years old. Together, the two Scouts spent their time at Camp Century helping with ice-core drilling, maintaining the camp, making weather measurements, assisting with the newly installed reactor, and drafting articles for magazines and newspapers about their stay.[23]

Days after he returned to the United States, Goering (along with two false Goerings) was the opening act on *To Tell the Truth*, a prime-time CBS television quiz show where multiple contestants present themselves as the same person, but only one of them is for real.[24] There, a young Johnny Carson and Betty White quizzed the three young men for almost nine minutes. None of the panelists picked the real Goering, but the Scouts' answers covered the whole "City under the Ice" experience. White's last question, asking why live in the ice, got the most telling answer: "The purpose of the experiment is of course to see whether man can survive in arctic conditions on a year-round basis with an eye in the future toward having a year-round scientific base."

Perhaps Camp Century's most public moment came on Sunday, January 15, 1961, when CBS aired an episode of *The Twentieth Century* filmed when Walter Cronkite visited Camp Century in 1960. At the time, no one in American life was more trusted than this newsman. The episode, "City under Ice," is a 24-minute documentary about the camp.[25] In grainy black and white, Cronkite ends the show by climbing aboard an Army helicopter and summarizing the military's challenges in the Arctic: "Men will stay behind here, in their city under the ice to continue man's battle against nature. Can he stop the crushing force of the ageless ice? These men here will seek the answer."

The military's ambitions went even farther than polar regions.

When interviewing Camp Century's lead project officer, Major Thomas C. Evans, Cronkite asks, "This barren arctic wasteland is not a great deal different than the surface of the moon in some respects, does your work have any application for man establishing a base on the moon?" Evans responds, "Well, it's certainly not an immediate objective but I think probably the problems here are somewhat comparable." Cronkite wasn't alone in his thinking. Charles Michael Daugherty's 1963 book about Camp Century closes with this sentence: "The scientists and engineers presently concerned with arctic development are well aware that their next job may be on the moon."[26]

Clearly, much of the Army's work in Greenland was an expression of American military, engineering, and scientific prowess at the height of the Cold War.[27] It was into this Cold War environment of presumed American military superiority that Camp Century was born. But the Scouts on television and Cronkite's reporting were something more subtle—a clever way to put the Arctic and the ingenuity of American scientists and engineers into the public eye. The camp fit the American ideal of the 1950s and early 1960s—that guts, hard work, and resourcefulness could accomplish anything.

It takes a lot of energy to power a camp on an ice sheet. In the 1950s, that energy came almost exclusively from diesel fuel.[28] Its distinctive pungent odor permeated the Greenland camps and surrounded the heavy swings pulled by the Army's massive diesel-powered D-8 snow tractors. One winter, the men living under the snow at Site 2 nearly ran out of fuel—only a cache stored over winter at the seasonal Camp Fistclench saved them when airdrops weren't possible. The Army estimated that fuel alone was about 70 percent of the cargo (by weight) hauled to Site 1, Site 2, and Camp Fistclench in the 1950s. Moving that much fuel across the ice sheet was a logistical challenge and was itself energy intensive.

It became clear that the military in Greenland couldn't keep operating like this much longer—it needed to find a different energy source. By the late 1950s, nuclear-powered Navy submarines were routinely cruising beneath the sea ice covering the North Pole, while the Army was dragging sleds filled with diesel fuel across the ice sheet and dropping fifty-five-gallon drums of fuel from airplanes. Thus, there was good reason and precedent to power Camp Century with a nuclear reactor. Forty-four pounds of uranium would take the place of a million gallons of fuel, and the foul taste, expense, and logistical difficulties of diesel would mostly be a memory.

The Army Nuclear Power Program (ANPP) began with the 1954 marriage between the Army Corps of Engineers and the Atomic Energy Commission.[29] ANPP focused on developing small nuclear reactors to support military missions, including providing power to remote Arctic radar installations, creating mobile atomic power stations on trucks and ships, and designing portable, modular power plants like the one eventually installed at Camp Century. In the 1950s and 1960s, the ANPP built and deployed eight reactors. None still operate, and decommissioning and cleanup have proved costly.

In just three years, engineers designed, built, and started the first Army reactor, known as the SM-1A (which, decoded, indicates that the reactor is Stationary and Medium power). The American Locomotive Company of Schenectady, New York (ALCO) built and then installed that reactor at Fort Belvoir, Virginia (just south of Washington, DC). In 1957, the SM-1A went critical, producing up to two megawatts of power. It was the first reactor to contribute electrons to America's power grid. Before the Army shut down the SM-1A in 1973, more than 800 Army nuclear engineers trained on the reactor, which powered the base and parts of northern Virginia.

In January 1959, the Army again selected ALCO when it came time to build Camp Century's PM-2A (Portable, Medium power) nuclear power plant. ALCO had to build, test, and ship the entire plant in modules by May 20, 1961.[30] Every module fit in a cargo aircraft, and ALCO engineers designed the reactor for rapid assembly in the field. The Army included increasingly severe fees for delayed delivery, making it clear that the atom would power Camp Century in a timely manner. The total cost (including shipping and installation) of nearly $5.7 million for the reactor was nearly two-thirds of Camp Century's construction budget.[31]

There may have been another reason for moving this project along so quickly. There's good evidence that a top secret, fantastical Cold War idea called Project Iceworm drove the U.S. Army to build Camp Century.[32] If the Army had gotten its way, six hundred American nuclear-tipped ballistic missiles would have been moving through tunnels on rails inside the Greenland Ice Sheet—something akin to an Arctic game of missile chess where the Soviets would have been hard pressed to guess the next move.[33] In the Army's view, the advantages were clear: "mobility with dispersion, concealment, and hardness. . . . Located in a barren wilderness far from American population centers, Iceworm would probably remain 'relatively invulnerable' for years."[34]

The project would have covered much of northern Greenland—52,000 square miles, the size of Louisiana—and there would have been sixty different under-ice missile launch control centers. Built in trenches like those excavated at Camp Century, the missile network and its 11,000 operators would be well shielded should a nuclear war break out. Nuclear reactors would power the system—diesel generators were insufficient. Camp Century was a test-bed for Iceworm in several ways—the trenching, the reactor, the barracks, the men's sanity, and the short rail line.

The Army took Iceworm seriously enough to solicit a detailed cost estimate for running a rail line from Thule to Tuto and then through the ice and snow to Camp Century, and eventually onward to "launching sites."[35] Travel time from Thule to Century in any weather conditions would be just a few hours. The price tag, without the add-on of nuclear-powered locomotives, which would minimize tunnel-warming heat, was $47.3 million in 1962—more than six times the cost of Camp Century. There were add-ons for bridging crevasses that the engineers knew would form, and for tunnel trimmers to remove the creeping ice and snow. Even with all this, the expected lifetime of the 160.7-mile-long railroad was years, not decades. Iceworm was both technically and financially untenable.

Camp Century's reactor came to Thule by ship, arriving in July 1960.[36] The cargo included 300 tons of reactor parts packed on skids ready to drag 138 miles across the ice. Cranes, trucked to Tuto, loaded the skids onto heavy swings. One specially made sled carried the steel vapor container, which was so heavy it took four massive D-8 tractors to drag it across the ice. To show that the reactor was indeed portable, a four-engine, double-decker cargo plane airlifted one of the reactor cooling units, weighting about 15 tons, to Thule.

When the PM-2A arrived at Century, the reactor trench was ready. Cranes lowered parts into place, steam fitters connected pipes, and electricians ran control cables while carpenters built the reactor housing and controls building. By fall, engineers had assembled the reactor. On October 1, they installed thirty-two uranium fuel rods in the reactor core, and testing began.[37] The uranium in the PM-2A fuel was highly enriched (93 ± 1 percent uranium-235), allowing for a small, compact reactor core. Natural uranium contains only 0.7 percent ^{235}U. This isotope is a critical ingredient for nuclear reactors because it sustains nuclear fission, splitting when hit by a neutron and at the same time generating more neutrons, which go on to split more uranium nuclei.

The PM-2A at Camp Century had two distinct sections. The first was the reactor, which held the core of nuclear fuel in a containment vessel. The fission reactions in the core—which define nuclear power—produced heat and a lot of radiation (primarily neutrons, but also more penetrating gamma rays). The second was the power plant, a steam turbine unit that spun a generator. This section was quite conventional and had the same design as a coal, oil, or gas power plant.

The engineers at Camp Century wanted to distribute heat, but not radiation, to the camp. In order to contain the radiation produced by the reactor, they had to isolate the water that cooled, and transferred heat from the reactor, from the water that generated power and heated the base.[38] Two separately piped loops, connected by a heat exchanger, did the trick. Nuclear reactions in the core warmed water in the primary cooling loop to 479°F, twice water's boiling point. But the water couldn't boil because the pressure in the reactor and the primary loop was too high, at 1,300 pounds per square inch. The water in the primary loop became radioactive as it circulated through the reactor. The water in the secondary loop, isolated from the reactor, did not. As the water in the secondary loop flashed to steam, it drove the turbine that spun the camp's generator. Another closed loop, filled with ethylene glycol antifreeze, condensed the steam after it passed through the turbine. The warm glycol was cooled using a sump in the ice sheet, where the waste heat melted what the Army estimated was 20 million gallons of water for use in the camp.[39]

At 6:52 a.m. on October 2, 1960, engineers began retracting the control rods from the reactor. These rods had soaked up the neutrons emitted by the radioactive uranium fuel, and thus—until that moment—had prevented a nuclear chain reaction. As the rods rose each fraction of an inch, more neutrons escaped, and the rate of fission increased until the reactor went critical, meaning that the nuclear reaction was steady and self-sustaining. But almost as soon as the reactor went critical, engineers shut it down because of a leak in the vapor container. That was a fixable problem. But, by taking the reactor critical, the engineers discovered that the first and only reactor installed in an ice sheet had a serious design flaw. The PM-2A was shedding far too many neutrons—endangering its operators and potentially the reactor itself. Camp Century's diesel generators started back up.

* * *

As soon as the reactor shut down, the head-scratching and redesign began. ALCO engineers in New York quickly crafted lead, steel, and boric acid shields—all designed to absorb wayward neutrons. On December 10, the new parts arrived at Camp Century, and their installation and testing began.

Oliver Gene Drummond was a career Army man. In 1960 and 1961, Drummond's orders sent him to Camp Century.[40] He spent months on the ice, separated by breaks of weeks to months. He'd come home bearing presents for his young daughter and then fly back north again. Later that decade, Drummond served in Vietnam and amassed more than a half-dozen medals—among them the Oak Leaf Cluster, the Bronze Star, the Gallantry Cross with Palm, the Armor Badge, and the Campaign Medal. Drummond's daughter, Dagmar Cole, knew her dad as a serious, no-nonsense man, never prone to exaggeration or tall tales.

Drummond came home to Oklahoma from Camp Century with stories.[41] He told his daughter about the eerie, brilliant blue glow of nuclear fuel rods bathed in water as he helped put them in the reactor before it started up. He was describing in perfect detail the Cherenkov radiation caused as electrons emitted by the radioactive uranium interacted with the water surrounding the fuel rods. Most striking to Cole were her dad's recollections about two different women who came to Camp Century, a veterinarian and an engineer. But as an enlisted man, he never learned their names.

The veterinarian was Danish. For most of the year, she went from village to village along the coast caring for the native Greenlanders' dog teams. She had a dog team and sled of her own. On one of her visits to Camp Century, the woman traded her dog whip for an insulated military hat with earflaps, but she refused to sleep in the warm camp under the snow. Instead, she dug a snow cave of her own, piled in with the dogs, and spent the night with her team to keep warm.

Why would a veterinarian drive her dog team and sled 138 miles over the ice sheet to visit Camp Century? Perhaps to vaccinate Mukluk. In the late 1950s, rabies infected the Arctic foxes of northwestern Greenland. Many of the animals died, but not before infecting sled dogs, more than 300 of which were ordered destroyed to prevent the infection from spreading.[42] The population of foxes around Thule had exploded, perhaps as they fed from the base dump. The solution Army cooks came up with was to destroy all leftovers. In 1960, the Thule Air Base vet and his Danish colleague waged an aggressive vaccination campaign.[43] It's possible that colleague was the woman who traveled to Camp Century.[44]

Then, there was the engineer who helped get the nuclear reactor running. Drummond's story was simple: A woman appeared on a helicopter from Thule late in the fall of 1960. She sat by herself at meals. She lived in officers' quarters because that was the only place available where she could live without men around her. During the day, she set up at a card table in the mess hall and did calculations with a jar full of pencils, pads of paper, and a slide rule. Drummond reported that the woman was confident and seemed to enjoy being alone.

On what he remembered as her third or fourth day in camp, she handed a to-do list to the reactor chief and demanded a helicopter to take her back to Thule. When the base commander ordered her to stay until the reactor worked, she told him she was a civilian and that he had no charge over her. She went to the radio operator, and by that afternoon, a helicopter arrived, and she was gone. Drummond recalled that the men performed the tasks on the list and started the reactor soon after.[45]

Five months after the PM-2A first went critical, the reactor went into service, and it ran reliably for more than two years. The diesel generators were always available as a backup if needed, but Camp Century's electricity was, until the summer of 1963, largely free of fossil fuels. Splitting atoms took the chill off the Arctic, warmed the undersnow barracks, lit the tunnels, and melted ice to provide water for the camp. Nuclear heat provided unlimited hot showers, something the Army proudly touted in movies and news reports.

The PM-2A produced 1.5 megawatts of electricity. The plant also produced steam used to melt ice for drinking water. That steam alone replaced the 65,000 gallons of diesel fuel used each year for the Rod well before the reactor went online. If the PM-2A were running today, it could power over 1,100 American homes. Camp Century, with at most 200 occupants, was an energy hog. Yet, it was this abundant power (both nuclear and diesel) that facilitated ice-core drilling over six years. Not only were the core drills electrically powered, but so were the barracks where the drilling team lived and the mess hall in which they ate.

For the nearly two and a half years that the PM-2A ran, neutrons continued to spew from the reactor at rates well above the design specifications. Those neutrons were irradiating the reactor and everything nearby. Such stray neutrons can cause a lot of trouble because these chargeless subatomic particles make everyday elements radioactive, and thus hazardous, through the process of neutron activation. Adding a pair of neutrons to the hydrogen in water makes tritium, which has two neutrons and one proton. Tritium is the most common radionuclide produced and released by nuclear reactors. With a half-life of 12.3 years, tritium emits beta particles. These particles have little mass or energy, and thus little ability to penetrate the skin. Tritium, unless ingested, presents only a low hazard.

Other activation products are more dangerous. Sea salt, blown onto the ice sheet by storms, found its way into the reactor's cooling water, which the Army obtained by melting ice and snow. Sea salt contains sodium-23, which becomes radioactive sodium-24 when irradiated by neutrons. Sodium-24 doesn't last long. It decays to radioactive magnesium with a half-life of 15 hours, emitting beta particles. But the magnesium emits gamma rays, which are far more dangerous because they carry a lot of energy, allowing them to penetrate tissue deeply and cause significant damage. Cobalt, which makes steel alloys stronger, forms a more troubling family of neutron-activated isotopes. Many are gamma-ray emitters, and the most common ones have short half-lives—hours to years—meaning that they emit a lot of damaging radiation quickly.

So many neutrons were escaping from the PM-2A that the Army took action to stop at least some of them before they activated materials outside the reactor. Men sprayed the icy walls of the reactor trench with a solution containing boron, one of the most efficient neutron absorbers. When boron absorbs a neutron, it splits and transforms into stable and harmless helium and lithium. Soldiers filled fifty-five-gallon steel drums with water, which soon froze. Walls made of these drums, stacked one on the next, lined the ends of the reactor tunnel, absorbing the neutrons that would otherwise have been bombarding men working nearby. There was a downside to these preventative measures, however. Neutrons spewing from the reactor were not only interacting with the ice, but also with the steel reactor building and the water-filled steel drums. Steel is made of iron alloyed with chromium and cobalt—all excellent neutron-activation targets.

Records of radiation exposure, albeit scarce, suggest that the Army's fix protected the men operating the reactor. The average radiation dose for a reactor engineer was 50 mrem/month of gamma radiation and 40 mrem/month of neutron radiation.[46] Reactor crews at Camp Century received an average dose of less than 300 mrem during each of their three-month rotations. The Army reported no personnel overexposures for reactor technicians. The men working around the Camp Century reactor remained well within safety limits for occupational radiation exposure at the time.[47] But, given that the average American absorbs about 300 mrem over an entire year, their exposure was significant in comparison.

Neutrons leaking from the reactor were not the only radioactivity released by PM-2A. Design drawings for the reactor show radioactively "hot" water from the primary cooling loop draining to a "hot waste" tank. From there, a pipe leads to the "hot waste disposal sump," a Rod well at the edge of the camp into which the radioac-

The reactor trench at Camp Century, January 1961. The hot waste tank is inside the metal building. Beyond is the wall of metal drums filled with water (ice) to soak up stray neutrons from the reactor. The Boy Scouts stand along the trench wall. *U.S. Army photograph.*

tive liquid flowed after being treated and discharged by the reactor engineers. Untreated primary cooling water is nasty, hazardous, and radioactive stuff. Metal in the reactor, including pipes and fuel rods, slowly dissolves, leaching into the water. Neutrons bombard the water (and every impurity in it) as it flows through the reactor, transforming dissolved elements like iron, cobalt, and manganese into their highly radioactive cousins.

Sometimes, the engineers needed to drain the primary cooling loop for maintenance. When that happened, protocol mandated that they filter the water and use chemical treatments to reduce the dissolved load—even so, the water remained somewhat radioactive. After leaving that water in the hot waste tank briefly so that very short-lived nuclides could decay, the engineers would discharge it into the hot waste disposal sump. The Army monitored the water's radioactivity as it poured into the ice sheet. Reports suggest that levels of radioactivity discharged remained within limits set by the Danish government.[48]

At Camp Century, radiation also fell from the sky with the snow—the result of atmospheric testing of nuclear weapons. The highest fallout rates at Camp Century occurred in 1962 and 1963, coincident with some of the most intensive atmospheric testing of thermonuclear weapons at Novaya Zemlya, the Soviet test site nearly 2,000 miles to the east.[49] Similar layers of radioactivity occurred across Greenland's near-surface snow, even reflecting American atomic bomb tests in the tropics.[50] The men at Camp Century weren't blind to this fallout, and they used Geiger counters to take radiation readings regularly. They needn't have worried about radiation in their drinking water, however, because that water came from snow that fell well before the first atomic bomb detonated.

View of the drill tower in Trench 12 at Camp Century. The hand-lettered sign behind the rig features a skull and crossbones and reads, "Danger. No! smoking around the pit," probably a warning reflecting the presence of flammable TCE and diesel fuel vapors from the drilling fluid. *U.S. Army photograph. Robert W. Gerdel Papers, SPEC.PA.56.0022, Byrd Polar and Climate Research Center Archival Program, Ohio State University, Box_2_Folder_29_015.*

HITTING BOTTOM

> This project is one that grips the imagination and creates enthusiasm. It is a challenge which will attract high scientific talent.
>
> —HENRI BADER, *Scope, Problems, and Potential Value of Deep Core Drilling in Ice Sheets*

Nearly a mile of ice separated Camp Century from the bed of the glacier below. When the Army built the camp, there was no technology capable of drilling all the way through the ice sheet and bringing intact ice cores back to the surface. The solid-stem drilling at Site 2 in 1956 and 1957 retrieved sections of core, but the process was slow and the core quality variable and often poor. The deepest core at Site 2 penetrated only the top third of the ice sheet—representing at most a few thousand years of history. For Bader, that wasn't enough. He argued strenuously in favor of coring to the bottom of the ice sheet, and laid out an ambitious plan for analyzing the resulting core—in a haunting way, anticipating almost exactly what scientists would do in the next six decades.[1]

In 1957, Bader directed B. Lyle Hansen, a U.S. Army engineer, to try a different approach: a thermal drill that would melt its way through the ice. Hansen hired American engineer Herb Ueda to work on the project, not because Ueda had experience with ice, but because the two men shared a background in farm work. Ueda was just out of school. For three years during World War II, the U.S. government had forcibly interned Ueda and his family in Idaho because they were of Japanese descent. By 1958, Ueda, still under thirty years old and father to four children, was employed full time by the Army Corps of Engineers.

Ueda recalled that Bader "was a pretty intelligent sort. He figured, well, if it's ice why can't we melt our way through, drilling a hole, you know. So, they applied for a grant from the U.S. National Science Foundation and got it."[2] An engineering company located near SIPRE's headquarters purpose-built the drill for this Arctic assignment. The Army was optimistic about the new drill, and so was the press. A journalist's analogy is descriptive: "The drill, operated by electrically heated elements, melts its way through the ice much the same as a hot soldering iron would pass through a barrel of butter."[3]

In 1961, *Science Digest* claimed, with no evidence except the Army's word, that the thermal drill would melt its way through 12,000 feet of glacial ice and retrieve five-inch cores from a six-inch hole.[4]

> Heated to a maximum of 400°F, it melts its way around a core of ice at a rate of six inches a minute. The core is held in the drill and removed when the drill is lifted. By successively removing ten-foot ice cores, the drill can reach its maximum penetration. . . . The thermal drill was originated to break the ice barrier which previously limited ice drilling to 1,300 feet . . . this research may provide clues to the history of previous climates.[5]

SIPRE scientists and engineers worked for several years to meet Bader's challenge. In July 1959, they drove the newly designed thermal drill to Camp Tuto, up the ice ramp, and a few miles onto the ice. After several months of setup and drilling, they penetrated only eighty-nine inches of ice—a little more than six feet. Undeterred, in September 1960, the men and the drill returned to the ice above Tuto and tried again. This time, they drilled forty feet into the ice, just enough that the entire drill, which was thirty feet long, was completely below the ice surface. That was sufficient success that the men and the 900-pound drill moved by heavy swing to Camp Century in October 1960. Soon they had installed the drill in the science trench. It hung from a thirty-foot tower on an inch-thick wire cable. The winch and the 12,000 feet of cable weighed nineteen tons. Like the Taku drill, this drill was a wireline design, so core retrieval was far more efficient than at Site 2.[6]

The drilling at Camp Century caught the attention of Albert Crary.[7] Between 1960 and 1968, Crary was the chief scientist for the U.S. National Science Foundation's Office of Polar Programs. That agency provided $600,000 to support the Camp Century coring and analysis effort. *Science Digest* noted that Crary "has a special fondness for the project" and quoted him as saying, "At last we will be able to read the layers of time from ice cores thousands of years old." Then he paused and added, "If we ever get to 10,000 feet, it staggers the imagination to think what we may find."[8]

Then, as now, collecting ice cores is difficult and expensive. It requires living on the ice far from civilization, operating machines in the bitter cold, and dealing with extreme weather. Camp Century provided a dramatic advantage over the previous ice-coring camps: the snow tunnels sheltered the drill trench, as well as all the men and their lodging, from the weather above. No matter the cold or wind on the surface, the air was still and the temperature moderate in the trench. There was consistent power for coring and for core-analysis equipment, including the light table used to identify layers in the ice. Despite these favorable conditions, the initial drilling at Camp Century didn't go well. The thermal drill malfunctioned and melted the ice core. After four months in Greenland, the ice-coring team departed Camp Century in December, disappointed. The first science project in the newly completed camp was off to a rough start.

By March 1961, the drillers returned to Camp Century with a rebuilt drill. That year, they collected a 186-meter core, but the drill

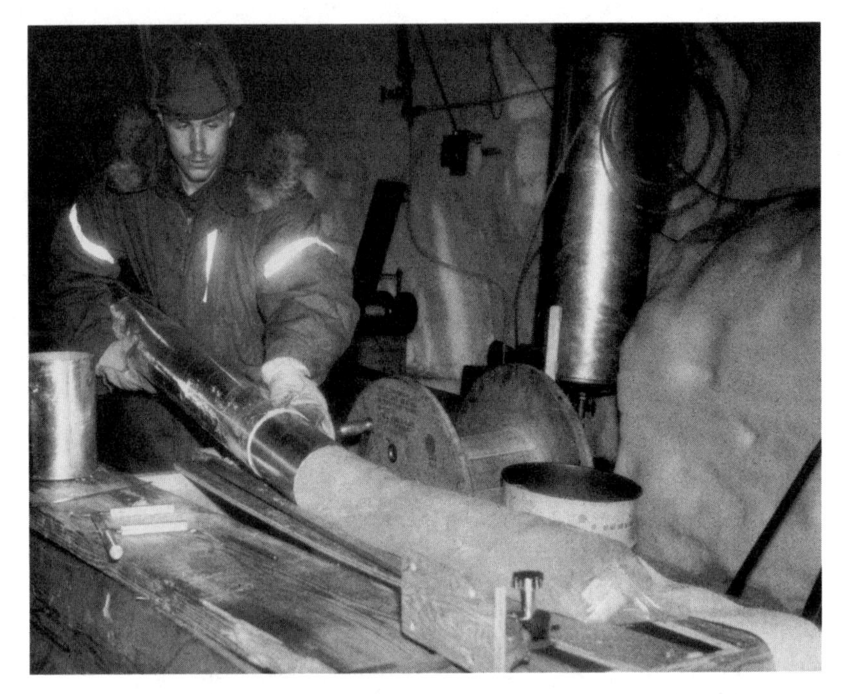

George C. Linkletter, a CRREL employee, laying a section of the Camp Century ice core on the light table for cleaning and logging in the drill trench. Undated, circa 1961–1966. *U.S. Army photograph, Specialist Fifth Class, Charles T. Bodey.*

got stuck at the bottom of the hole, and the team had no choice but to abandon that drill and the hole and move on. Their next hole, drilled in 1962, produced a 780-foot core, but this time the heating head detached from the drill, and again they had to move the drilling rig and start over.[9] A third hole, drilled after they procured a new heating head, produced 866 feet of core. At the end of the summer 1963 drilling season, the camp, now without a working reactor, shut down for the winter. The Army filled all three holes with drilling fluid for the first time so as to help keep them open when there was no drilling. Coming back in the summer of 1964, the team continued drilling the third hole. They drilled with fluid to a depth of 1,755 feet, collecting core and setting a new depth record. But it was yet another failed attempt. The drilling fluid destroyed the insulation on the core cable and left a mess in the bottom of the hole.

Glen's flow law was haunting the ice-core drillers. Cut a hole in an ice sheet, and eventually the hole will close as ice around that hole deforms. That ice flow is the reason drilling fluid fills every deep core hole. Without fluid, a mile-deep hole, about the width of a man's hand and years in the making, would eventually squeeze closed under the crushing pressure of Greenland's massive but slowly flowing ice. Watery drilling mud, the standard drilling fluid where it's warmer, fails in Greenland, solidifying in the cold because the ice and firn are below the freezing point of water.

The drilling fluid used at Camp Century was a mixture of diesel fuel (87 percent) and trichloroethylene (13 percent). Trichloroethylene (TCE) kept the thick diesel liquid and flowing in the bitter Arctic chill. This organic solvent, invented in 1864, takes caffeine out of coffee, extracts oil from soybeans, and removes grease from clothes. It's nasty stuff, probably carcinogenic to the kidney, liver, and lymph systems after prolonged exposure. Shorter exposure leads to intoxication, heart irregularities, dizziness, and confusion. Until banned for human use in 1977, TCE was a workhorse anesthetic. Doctors used it for women during childbirth, although it ended up leaving a few with facial paralysis as a side effect.[10]

As the Camp Century borehole penetrated deeper into the ice, the drillers added more fluid. There's no record the fluid ever came out of the hole—today, 7,000 gallons of toxic organic liquid sit in the ice.[11] That's more than enough contamination to warrant a major cleanup effort stateside, which might start with pumping out the liquid "product." If the hole penetrated soil or rock rather than snow, engineers would use "pump and treat" methods, along with air stripping or activated carbon absorption, to reduce the trichloroethylene concentration in the groundwater.[12] And they might address any remaining diesel fuel using bioremediation, which relies on bacteria that eat petroleum products. Neither is a good option on the ice, where there's no liquid water and little life.

OVER THE SIX years it was open, scientists flocked to Camp Century. They saw the camp and its unique setting as a place to conduct research on many topics, including the behavior of the solar system and the history of Earth, as well as a place to measure the impact of people on the planet's environment.

Some came to find evidence of the Cold War. Atmospheric testing of nuclear weapons, starting in 1945, sent radioactive isotopes into the atmosphere that nature rarely or never did.[13] These isotopes were by-products of nuclear fission and fusion weapons. The most conspicuous included the following:

Americium-241, 432-year half-life, concentrates in bone, liver, and muscle
Cesium-137, 30-year half-life, distributed throughout the body, behaves like potassium
Iodine-131, 8-day half-life, accumulates in the thyroid
Strontium-90, 29-year half-life, accumulates in bone, behaves like calcium
Tritium-3, 12-year half-life, incorporated in water molecules
Plutonium-239, 24,100-year half-life, inhalation risk

Atomic blasts in the atmosphere released these nuclides, and some of that fallout settled onto Greenland's snow. American and Soviet scientists designed their bombs differently, so the isotopes of plutonium each generated were unique—an international fingerprint.[14] Langway and his team tracked this radiological contamination, sampling snow pits, counting layers, and thus dating when and how much plutonium fell from the sky onto Greenland.[15]

Others saw an opportunity to study the heavens using dust that landed on the ice sheet and eventually ended up within the ice.

Camp Century used 3.7 million gallons of water a year (that's 10,000 gallons each day).[16] By running hundreds of thousands of gallons of that water through filters capable of trapping small bits of cosmic dust that fell to Earth, cosmochemists collected micrometeorites—the largest were the width of a human hair. They both identified the material in the micrometeorites and tried to measure rare isotopes made by cosmic rays. Theirs was a heroic effort, yielding only a few ounces of material. Given the low sensitivity of analytical equipment in the 1960s, the team did science that would otherwise have been totally impractical.[17]

In 1964, Willi Dansgaard, the Danish physicist who had measured stable oxygen isotopes in rainfall using a beer bottle in his backyard, came to Camp Century. Accompanying Dansgaard on this trip was Henrik Clausen, a chemical engineer and Dansgaard's assistant at the time.[18] The two were in Greenland to collect tons of snow from the deep, inclined shaft that the Army had cut into the ice sheet below Camp Century. Their goal was dating ice, but their experiment wasn't a simple one. The protocol included melting tons of snow and then precipitating the water in a soup of chemicals hauled across the ice sheet. Dansgaard, with his interest in oxygen isotopes, wanted to see the drilling operation, but the drill trench was off limits. His diary entry for Friday, July 31, 1964, reads,

> The drilling goes on night and day, but there is no access for unauthorized persons. We were dissuaded to try because the drill master is a very determined gentleman. Sounds as if some military secrets are involved. What a shame because the ice core must be very interesting for measuring stable isotopes. What the Americans are going to do with the ice core is unknown. We might have referred to the rule that foreign expeditions collecting material on Danish ground are under an obligation to share it with Danish scientists, if need be. However, the way we got access to the ice core later on made things develop more harmoniously.[19]

Claire Patterson, a geochemist at Caltech, also got access to do science at Camp Century. The research he did in northwestern Greenland eventually had a great impact on the health of people around the world. Patterson got his start dating meteorites, which relies on precise measurements of the isotopes of uranium and lead. He began by measuring the amount of uranium. Although these measurements were not simple, they were doable, and the data reproduced well. In each meteorite sample, he then tried to measure the amount of lead that had accumulated as the uranium decayed radioactively over billions of years. The problem was, Patterson kept measuring too much lead—so much extra lead that he could not date anything, or at least could not date it correctly. Patterson concluded that inadvertent lead contamination was causing the problem with his measurements.

Looking back, it's easy to see why. Patterson was trying to analyze lead in the 1950s and 1960s, when the element was everywhere—in paint, in gasoline, and in people's bodies, their hair, their skin, their bones, and their brains. His solution was creating a new kind of lab and then going on an extraordinary cleaning binge. By doing all of this, Patterson created and maintained the first clean lab, of which the lab we use today to date samples from Greenland is a direct descendant. Patterson's protocol mandated that he

> distill all acids in special stills,
> triple-clean lab glassware with strong caustics,
> wear a full bodysuit in the lab, including a hairnet,
> cover the lab in plastic sheeting,
> wash the floors and the walls over and over,
> pump in purified air, and
> use no lead solder in the lab electrical system.[20]

The cleanup worked. In a measurement that has stood the test of time, Patterson showed in 1956 that the age of the meteorites, and thus of Earth, was 4.55 ± 0.07 billion years.[21] One might think establishing the age of Earth was fame enough for a young scientist, but it led Patterson down a path that turned out to be far more important. He spent the rest of his career making measurements that proved lead levels in people and the environment were far higher than they were when humans evolved.

To implicate human action in the rise of environmental lead levels, Patterson and his team needed to measure background levels of lead in the environment reliably. He started by collecting and analyzing old water from deep in the Pacific Ocean and bones from 1,600-year-old Peruvian mummies. In search of ice that he hoped would tell him how lead levels in snow (and thus the atmosphere) had changed over thousands of years, Patterson went to Greenland and spent time in the snow and ice tunnels of Camp Century and Camp Tuto.

In Greenland, Patterson collected three sets of samples: ice from the clear blue walls of the Tuto ice tunnel; firn from an inclined shaft cut, on a twenty-degree pitch, to a depth of 300 feet below Camp Century; and fresh snow from a site fifty miles inland from Camp Century where no vehicle had ever gone (of course, he sampled upwind of his own vehicle).[22] His goal was to determine background levels of lead before people had started smelting large amounts of ore that contained both lead and silver (more than 3,000 years ago), and before Thomas Midgely's 1920s discovery that adding toxic tetraethyl lead to gasoline made internal combustion engines run more smoothly. It wasn't a good omen that Midgely had spent months in bed, severely poisoned by the organic-bound, and thus readily bioavailable, lead of his own invention.

Patterson's sampling recipe in Greenland was obsessive—a necessity if he were to measure the actual lead levels in his snow and ice samples instead of modern contamination. His team avoided sampling ice in Camp Century's main tunnels because vehicles using lead fuel had contaminated the snow there. Before sampling more remote tunnels, they chipped away six inches of surface ice from the tunnel walls. The team had acid-cleaned everything that would touch the samples and handled the samples with tongs, placing them into plastic-lined drums.

At Camp Century, people were Patterson's worst enemy because they were full of lead. He complained that

> Army engineers who excavated the shaft smoked profusely and consumed large quantities of carbonated beverages, which came from cans that had been sealed with lead solder. These persons were consuming and excreting large quantities of industrial lead. In addition to blowing their noses and spitting, they also urinated on the walls and floors of the shaft. The concentration of industrial lead in these body excretions were about 100,000 times greater than the natural concentrations. . . . A single fiber or particle of hair, skin, or clothing could readily contribute more industrial lead than would be contained in all the firn or ice collected from one sample.[23]

Paranoia paid off. Although Patterson's analyses showed that Camp Century sewage had a million times more lead than the old snow and ice surrounding the disposal well, the 2,800-year-old ice from Tuto's tunnel was unleaded. By 1753, in the deepest Camp Century ice sample he measured, lead concentration had risen twentyfold. Blame that on the Industrial Revolution. Lead concentrations in the snow steadily rose—threefold more by 1815, doubling again by 1933, and then, in the 1940s, skyrocketing along with the widespread use of lead in gasoline. The snow at Camp Century in the 1960s had 500 times more lead than the old ice at Tuto. By 1969, when Patterson's paper came out, Midgely was long dead. Another one of his chemical inventions, the first chlorofluorocarbon gas, was by then wreaking its own havoc, destroying Earth's ozone layer.

Patterson showed that tiny bits of lead traveled to Greenland on the wind. The lead fell with the snow that, over decades, became ice. Today, thanks to sensitive instrumentation and well-dated ice cores, the outline of lead use that Patterson first created now exists in exquisite detail.[24] Ice from the upper few hundred meters of the Greenland Ice Core Project (GRIP) core, from central Greenland, speaks of the demand for valuable metals in Europe as civilizations rose and fell. Lead levels in ice cores rose along with Phoenician ascendance (peaking at 800 BCE) and fell as the Roman Republic began to crumble (100 BCE). Political stability meant more mining, smelting, and lead emissions. Epidemics like the Antonine plague, which was probably smallpox and struck between 165 and 193 CE, meant less demand for smelting. Ice from those years in Greenland contained little lead. The effects of plagues and political disruptions on lead emissions lasted hundreds of years. Thus, Greenland's ice is telling us that European societies, before the Middle Ages, were not very resilient or stable.

IN LATE SUMMER 1963, U.S. Army Commanding Officer Joe Franklin stepped off a military transport plane and into Thule's chilly Greenland air. He had arrived with a team of Army engineers to remove and relocate the reactor installed at Camp Century three years before. After running since late fall 1960, the PM-2A was now a radioactively hot reactor. Franklin had come to Greenland with no experience in reactor decommissioning. Actually, there wasn't a person on the planet with such experience. No one had ever decommissioned a used nuclear reactor.

While the Army suggested publicly that the reactor had completed its mission at Camp Century and was ready for reuse elsewhere, the truth appears to be more complicated. Neutron bombardment of the reactor containment vessel during power generation was known to weaken steel—but how much, and how quickly? This question had the Navy, with its growing fleet of nuclear-powered ships and submarines, alarmed. Email chatter between Franklin and other veterans from his West Point class of 1955 suggests that the reactor was moved from Camp Century for testing of the containment vessel. If the vessel remained strong, then the nuclear Navy would keep sailing.[25] If not, the Navy would call the ships and submarines back to port for repairs and refitting of their power plants, leaving a major weakness in the national defense.

Franklin's men started by removing the reactor's nuclear fuel—a challenging and dangerous task. Both the fuel and the reactor were highly radioactive, even though Camp Century's plant operators had shut down the reactor in July in preparation for the team. To remove the fuel, the engineers built a steel bridge and installed a large crane over the sixty-foot-deep snow trench that housed the reactor. Then, the team moved snow and cut hatches into the reactor building. They lowered seven lead-lined casks down to the reactor level and placed them underwater in the spent-fuel tank, where water would absorb most of the radiation.

Early on September 27, Franklin's team began retrieving the highly radioactive fuel rods—each consisting of enriched uranium in a stainless steel–clad tube—from the reactor core.[26] The core wasn't large; an average adult could wrap their arms at least halfway round it. The men removed the fuel tube by tube, using a long-handled extension rod, and loaded each tube into one of the casks, which were still underwater. As each cask filled, the crane lifted it from the tank, and the men cleaned the outside, purged the radioactive water, and winched the cask to the surface. They finished in little more than a day. Radiation levels around the casks were low, except at the ventilation ducts, where the radioactive air mixed with thirty-knot winds at −20°F. The men tried to avoid the ducts, where a Geiger counter read 3.5 roentgens per hour—enough radiation to sicken a person exposed for less than a day.

Removal of the PM-2A resumed in spring 1964.[27] A team, again with Franklin at its helm, removed the snow covering the reactor trench and fashioned a sloping ramp from the depths of the trench to the ice sheet's surface. After removal of some failed metal roofing, reactor disassembly began. The men then disconnected and packed the pipes, valves, and gauges. Even though Franklin's men

In spring 1963, heavy swings transported lead-lined casks containing uranium fuel rods from the PM-2A nuclear reactor over the ice sheet to Thule. *U.S. Army photograph, Private First Class Jon W. Fresch, May 13, 1964.*

had emptied the reactor core seven months earlier, radiation levels around the reactor remained dangerously high. Franklin measured more than 1,000 roentgens per hour, enough to kill most men outright if they lingered.[28] Extra lead shielding arrived so that the disassembled reactor could be safely palleted, then dragged on sleds over the ice for shipment back to the United States. Franklin, as commanding officer, insisted on doing the jobs that required the highest radiation exposure himself.

In less than three months, Franklin's team disassembled the reactor, transported it off the ice, and headed home. Back in the States, the PM-2A became an orphan. No other military camp wanted it, perhaps because the results of the reactor's nasty neutron-shedding habit were becoming all too clear. Most of the residual radioactivity resulted from the decay of cobalt-60 and manganese-54. Stray neutrons produced both of these isotopes by activating the steel that made up the reactor, its piping, and the containment vessel. Both isotopes emit dangerous and penetrating gamma rays, but at least they're short-lived.[29] Cobalt-60 has a half-life of 5 years; manganese-54, 312 days. By the time President Nixon resigned in 1974, the manganese-54 was gone. Today, more than ten half-lives later, the cobalt-60 is gone.

Two years after the PM-2A left the ice, and just after Camp Century closed for good, the reactor met its end in Idaho at the National Reactor Testing Facility. In 1966, the ten-by-three-foot containment vessel finally cracked. By design, the tank could withstand at least 1,750 psi. Destroying the reactor took 4,475 psi of pressure and a temperature of −20°F, conditions far harsher than it ever experienced at Camp Century. It failed only after engineers cut a gash in the flank of the containment vessel and dosed the crack with acid.[30] The Navy's reactor containments were safe.

There was a cost to the hasty removal of the PM-2A. In an email to his West Point classmates, Franklin casually relates the unwelcome news. Describing a trip he took soon after returning from Greenland, he writes, "The only thing I did which Lee escaped was to set off the whole-body-counter at Bethesda Hospital [the naval hospital outside DC] when I finally returned."[31] In the 1960s, that hospital housed a unique facility, funded by the Navy, that treated "victims of accidental exposure to radiation."[32] Franklin was radioactive, and that radiation was in, not on, him—a dangerous result, but perhaps an expected risk of decommissioning a nuclear power plant with no experience and limited protective gear.

How did Franklin become radioactive? When he arrived at Camp Century, engineers had already shut down the reactor, so there were few neutrons to activate the sodium, calcium, or iron in his body. Franklin likely ingested radioactive material. With the PM-2A having shed so many neutrons, everything in the reactor trench would have been at least slightly radioactive. Perhaps it was the bits of metal on his hands, or dust he inhaled from the air as the men took the reactor apart. Nuclear fission leaves behind a collection of unwelcome radioactive by-products.

The most highly radioactive material left Camp Century when the Army removed the reactor. But it's less clear how much of the dispersed, neutron-activated radioactivity the Army left behind when it abandoned the camp. Records indicate that radioactive cooling water flowed, as permitted, into the hot waste disposal sump on the ice sheet. Some radioactivity lingered at least in the snow, the metal reactor buildings, and the hundreds of fifty-five-gallon steel fuel drums repurposed as ice-filled neutron absorbers to shield those working in the reactor trench. There's no record that the Army removed any of this material along with the reactor, yet high numbers of neutrons bombarded all of it when the reactor was running.

The real-world results of Glen's flow law, which states that ice and snow abhor a void because they behave like a plastic material, were now threatening to shut down the rest of Camp Century. Everywhere, tunnel walls crept inward as the snow deformed, but the camp's reactor trench closed the most quickly. That trench was the warmest part of the camp, and the surrounding ice reacted accordingly. The trench containing the kitchen and mess hall, another of the warmest places in the camp, soon suffered the same fate. Here were empirical data validating Glen's flow law: warm ice was flowing much faster than cold ice.

The nuclear engineers had considered this issue, and their calculations were worrying enough that they acted. Gamma rays emitted from the core when the reactor was running were sufficiently plentiful to warm the snow beneath. The engineers added more shielding to prevent a whole different type of nuclear meltdown than affected Chernobyl, Fukushima, or Three Mile Island: the thawing of the foundation below the reactor. Still, the trench walls kept closing in.

To combat the deforming ice sheet, the Army dispatched soldiers with shovels and chainsaws to trim back the snow before it crushed the wooden buildings nestled in the narrowing trenches. To keep Camp Century open, men removed 20,000 cubic yards of snow from the tunnel system every year.[33] This was hard, manual work. Crawford Henderson, an Army engineer, had another idea. He designed a mechanical snow trimmer that ran along tracks on the sides of the tunnels, shaving snow, which it then crushed. The device used compressed air and blew the snow out of the tunnel in pipes.

In the early 1960s, Henderson traveled to Camp Century to test his invention. His wife remembers that during his visit, Henderson was the only Black man at the camp. Henderson received a regional Outstanding Young Engineer award from the Army in 1963 and authored an illustrated article for *Military Engineer*.[34] In the photo the Army published of the machine working in one of Camp Century's tunnels, a white soldier guides Henderson's invention along the snow wall.[35]

By the mid-1960s, the Camp Century experiment was drawing to a close. The camp's unique power source was dead and gone, and the camp was running on diesel generators. Project Iceworm never happened. Franklin was teaching nuclear engineering at West Point, the Vietnam war was raging, and the struggle for Black civil rights was in the nation's news magazines. In 1965 and 1966, there was one remaining scientific objective at Century: coring to the bottom of the ice sheet. With the camp open only in the summer by then, time was running out.

ARMAIS ARUTUNOFF WAS a Russian inventor and entrepreneur with dozens of patents. He immigrated to the United States in 1923 and founded the Reda Company in Bartlesville, Oklahoma, in 1930.[36] The company made some of the first successful and widely used submersible motors and pumps for the oil industry. Employing this expertise, Arutunoff tried his hand at building a drill for the oil patch. After nearly four years, the U.S. Patent Office approved Arutunoff's application for the Electrodrill—a wireline design in which a heavy cable that supplied power to the motorized bit replaced the solid string of drill rods.[37] A pump moved drilling fluid and rock chips away from the bit and out of the hole.

The Electrodrill was a large apparatus weighing nearly 900 pounds, with an 83-foot-long drill head.[38] Although it penetrated more than 1,200 feet into sedimentary rock when tested, its cutting rates were slow, and the drill had a habit of getting stuck.[39] When the Army came looking, Arutunoff was more than willing to sell his drill for cheap. For $10,000, the Army bought a reconditioned Electrodrill and modified it to drill through ice.

The new drill made its debut at Camp Century in 1965. It could collect up to twenty feet of core at a time. The ice cores were about four and a half inches wide. Glycol cooled the motor and melted the waste ice chips so that water, not chips, left the hole. The drill worked well enough that by the end of the 1965 drilling season, the depth of the hole at Camp Century had nearly doubled to over 3,000 feet—yet another ice-coring record. Perhaps the bottom of the ice sheet, still beneath more than a thousand feet of ice, was within reach before the trench walls closed in and the Army abandoned the undersnow base.

The coring team returned in the summer of 1966 and made rapid progress.[40] For example, between June 21 and July 1, according to the original handwritten drilling logs, the men cored nearly 800 feet of ice—almost 70 feet per day.[41] On July 2, those same logs report the first pebbles—the drill had entered the basal ice zone after coring 4,454 feet of clear ice. The men kept drilling. Four more times they pulled the core barrel out of the hole, cut the core into 5-foot sections of increasingly sediment-laden ice, and placed each section in a plastic bag and then into a numbered tube. Halfway down the next core barrel, the drillers' notes read, "30 cm [one foot] from the top silty ice stops and frozen sand begins." At that moment, the scientists accomplished the feat that had so long escaped them. Below 45 feet of silty ice and nearly a mile beneath the snowy surface, they had reached the bottom of the Greenland Ice Sheet.

The ice temperature at the bottom was a chilly −9°F. As a result, the ice was solidly frozen to the bed below. The men pressed on, drilling into the material under the ice and recovering over eleven feet of rocks, sand, and ice: frozen glacial till, some of which the log calls "permacrete." They placed the material sequentially in five different tubes, numbered 1059–1063.[42] Ueda notes that "only a worn bearing in the gear section prevented further penetration."[43] At least Hansen and Ueda were in the drill trench that day. Ueda describes completion of the core as "probably the most satisfying moment of my life, or of my career."[44]

On July 2, 1966, after almost six years of work, the Camp Century core was complete—a fitting and symbolic end to the U.S. scientific domination of the Greenland Ice Sheet that started in World War II. The men packed the drill and shipped it to Antarctica. They sent the last sections of the Camp Century ice core back to New Hampshire and the freezers at the Army's Cold Regions Research and Engineering Laboratory (CRREL), except for one rock, which Ueda held onto for a few years before donating it to a museum in Japan.[45] Unlike the drilling log, newspapers at the time, Langway's first paper, and Ueda's oral history suggest that the coring finished on America's birthday, July 4.[46]

ON SEPTEMBER 20, 1966, pieces of the Camp Century ice core made an appearance in an unexpected place: in glasses of Coca-Cola sipped at a Pentagon press conference. The ice traveled from CRREL freezers in New Hampshire to Washington, DC, in a cooler, probably packed with dry ice. A *New York Times* article reporting on the affair began by asking readers what they might expect soda cooled with ancient glacial ice to taste like.[47] As Langway and Hansen served up ice from the Camp Century core in paper cups filled

B. Lyle Hansen, Chester Langway, and Jay Thomas (*left to right*) holding two pieces of the Camp Century ice core at a Pentagon press conference on September 20, 1966. In the lower right corner, a stack of paper cups, presumably to hold the Coke served over glacial ice, sits just below a microphone, positioned to amplify the sound of the air bubbles in the ice popping. *Associated Press photograph by John Rous.*

with Coca-Cola, a microphone positioned over a cup broadcast the sound of ancient air bubbles bursting as ice from the time of Christ's death melted in the warm soft drink.

This event was a polar science publicity stunt, and perhaps a way of thanking the Army for a decade of logistical support. It was also an effort by scientists to put the science and engineering of ice coring front and center in the public eye. It worked. The Army and the Camp Century ice core got lots of press in the fall of 1966.[48] Although I wonder about the bouquet of drilling fluid in those drinks. Diesel, glycol, and trichloroethylene coated every segment of the ice core. Maybe Coca-Cola was stronger in the 1960s, or the palettes of the military brass less discerning.

Camp Century and its science had excited the media and the public one last time. That summer, the Army had abandoned the camp, and as fall came on and the days shortened, snowdrifts filled the sloping ice ramp that led down to Main Street. This winter was different than any since 1959. The next spring—and the next and the next—the ramp would remain uncleared as the camp disappeared beneath the snows that accumulated, on average, about six feet every year—enough snow to bury a soldier standing at attention. The reactor was gone, but almost everything else remained: the drill tower, the casing that lined the upper part of the borehole, the sewage and ethylene glycol coolant sumps, the barracks, the mess hall, the dozen toilets in two neat, back-to-back rows, and, perhaps, the pinups.

Eight days after the Pentagon press conference, Lucybelle Bledsoe died alone as a smoky fire consumed her apartment. Only in death did her name appear in a scientific journal, when Bender wrote her obituary in the *Journal of Glaciology*.[49] Her likeness gazes off the page in a black-and-white, head-and-shoulders portrait: narrow lips, cat-eye glasses, short hair, large earrings, and an inscrutable expression. Bledsoe's work and her life warranted nearly a page of text. Few glaciologists get such a posthumous honor in their field's journal, and never in 1966 a woman with an English major. Bender's postscript to the obituary is a stunning revelation about Bledsoe's place and influence in the science of Greenland that led to the Camp Century ice core.

> Too often we, as authors of scientific reports, take for granted the careful work of good technical editors and seldom do we give them the proper credit which they so rightly deserve. So it was with Lucybelle Bledsoe. Although you will not find her name listed as author or co-author of any research papers, she

has been a full-time worker in the field of glaciology for the last ten years.

After editing the papers and reports that made living and working in the Greenland Ice Sheet possible, Bledsoe, a woman in a man's world of ice and snow, never got the chance to know the secrets revealed by Camp Century's ice core.

For years, the field of ice-core science remained an "old boys club." But Bader, who had left Europe as the Nazi invasion loomed, was an exception, an early catalyst for change. Bader was deeply supportive of Wasserburg as a Jew and of Bledsoe and several other women at SIPRE as lesbians. He hired Black engineers at SIPRE.[50] This was a time in America when homophobia, racism, and anti-Semitism ran rampant. What was it in Bader's past that gave him such empathy for those with less privilege? Perhaps time and others will be able to answer this question. I wish that I could.

THE VIEW BACKWARD isn't the same as the view forward. The enthusiasm and platitudes the press and authors first lavished on Camp Century met reality in an unusual place: a book for young adults by Lee David Hamilton titled *Century: Secret City of the Snows*.[51] The title of the last chapter is "Health and Sometimes Happiness." Hamilton begins bluntly: "The psychological problems of living in an icecap city are many and complex. They are so complex, in fact, that no one is asked to stay at Century for longer than six consecutive months." Then, he goes on to list all the reasons why the city's occupants might not be happy:

insomnia and headaches
stuffy, smelly, and windowless buildings
toilet and shower facilities in a far-off tunnel
eighty-hour work weeks
nothing grows

> there is nothing to see
> the feeling of isolation, of living on another planet, is
> beyond belief

Hamilton finishes with a story that I've not read anywhere else. In the first year Camp Century was operating, three different men decided they'd be better off somewhere else and walked out of the main tunnel and onto the ice sheet. A rescue team took the men back to the base hospital, "poor and pathetic sights, all but senseless."[52] Then, there's the paper-clip soldier. He tells Hamilton,

> "If there weren't any paper clips around here, I'd go nuts!" . . . His paper clip chain was an ingenious device and the GI's pride and joy. Days of the month were represented by single clips, each strung together in an orderly fashion with a paper tag between every 30 or 31. On one tag, this radio operator had written *92 days until liberation*; on another, *62 days until the Messiah arrives.* Every morning when he arrived at work, off came a paper clip from the great chain. In this manner he kept track of the days and the months until, as he said, "The time of this white hell has ended for me."[53]

Danish Scout Soren Gregersen's diary is also revealing.[54] The happy Boy Scouts were a myth, a fallacy of the press. Soon after he'd come to Century, Gregersen wrote, "This is a horrible, desolate place. At least that's the general opinion and the daily topic of conversation among the Americans here." Gregersen described his time at Century with phrases like "bored stiff" and "quite an unfruitful day." Ueda's assessment of the Scouts' life at Century is similarly revealing.[55] "So, they had a little contest in Denmark and the United States. And the winner of these contests would get to spend a year at Camp Century. . . . So, the joke up there among the guys was, if Camp Century was first prize, what the hell did the runner-up get, purgatory? But, they survived."

The Scouts didn't have a mission at Camp Century. They spent a couple of days with the reactor crew until the reactor emitted too much radiation at startup. They measured snow tunnel deformation—a major problem already in the year-old camp. For a few days, they helped out with the drilling until the Electrodrill broke and the drillers headed back to the States. Then, the Scouts moved back to the reactor, making sawdust there for use as a neutron absorber in a failed attempt to lower radiation levels. Life was tedious, and by the end of their nearly six-month stay, the Scouts were ready to leave Camp Century.

The disconnect between men's accounts of their time on the ice and that of the American press is dramatic. They tell two quite different stories of the same place at the same time. It's a wonder, and a testament to all the men, and the drilling team in particular, that the Camp Century ice-coring effort, in its sixth year, finally made it to the bottom of the ice sheet. Drilling ice cores in the 1960s was like going to the moon—even the best and brightest scientists had only a hint of what they might find, but they knew that whatever they found, it was going to be interesting. What they didn't know is that the permacrete they worked six years to collect would more than five decades later change our understanding of the fragility of Greenland's ice, and thus of our planet's vulnerability to warming.

B. Lyle Hansen (*left*) and Chet Langway (*right*) examine the permacrete core from below Camp Century. In the foreground is the top of a core of basal ice from near the bottom of the ice sheet, displaying characteristic bands of silt and sand. Photograph undated, probably 1966. *US Army photograph, David Atwood, courtesy AIP Emilio Segrè Visual Archives.*

AGING

Science is not the result of dispassionate machines spitting out Truth; it involves passionate humans pursuing truth and fame and next week's paycheck, while satisfying curiosity at the same time.

—RICHARD B. ALLEY, *Earth: The Operators' Manual*

In May 1969, a team of CRREL scientists, along with several Danes, returned to Camp Century to document changes in the geometry of the borehole and the trenches. In the three years since the Army had left, snow had covered every trace of the camp at the surface except the weather tower, some vents, and a few escape hatches. After a ski-equipped C-130 transport plane dropped the men on the ice, they set up temporary surface lodging in a snowstorm. The team explored as much of the aging undersnow camp as they could access, documenting it all with photographs.[1]

The camp was a mess. Creeping snow had narrowed the trenches, crushing many of the buildings. Roofs had collapsed. Snow deformation had bent the pipes along Main Street. The heaving floor in the trench meant to demonstrate the feasibility of Project Iceworm had warped the empty rails. Abandoned steel drums sat in frost-covered rooms. Still, parts of the camp remained unchanged—the men were particularly grateful for the "tons of deep-frozen beef and thousands of prepacked rations of delicacies" that the Army had left behind when they shuttered Camp Century at the end of summer 1966.[2]

The men found the drill trench and the abandoned borehole.[3] They extended the metal drill casing ten feet above the snow surface to mark the hole's location and allow for future down-hole measurements, even after the camp and the trench were no longer accessible. The team slowly lowered sensors nearly a mile down the borehole and measured the temperature of the drilling fluid, which, having sat undisturbed for three years, reflected the temperature of the ice sheet at each depth. A second instrument measured the deformation of the hole, which indicated the flow of Greenland's ice sheet in the time since the drilling team cut through the ice.

Since the Army abandoned the camp, the flowing ice sheet had carried the borehole and the casing several tens of feet to the southeast and toward the coast.[4] Ice at the base of the hole had deformed so much in three years that as the men lowered the instrument down through the drilling fluid, it got stuck near the bottom. They never reached the deepest part of the hole or the base of the ice sheet. After two weeks, the plane returned to retrieve the team, who gave the aircrew a tour of the abandoned camp. Then they all climbed out of the snow and onto the plane, the last people to ever set foot in Camp Century.

Six months later, in October 1969, Dansgaard, along with Langway, published the first results from the Camp Century ice core in the journal *Science*, in a paper titled "One Thousand Centuries of Climatic Record from Camp Century on the Greenland Ice Sheet."[5] Their analysis of stable oxygen isotopes in the layers of ice revealed a 100,000-year record of Earth's warming and cooling—something no one had ever seen or done before. Dansgaard's research fundamentally changed how scientists and the scientific community thought about ice cores. It showed for the first time that glacial ice preserves a continuous and detailed record of past climates stretching back far beyond recorded history. Almost 20 years after Anderson pulled the first American ice core from the Taku Glacier,

Dansgaard's data confirmed Bader's vision of what was possible and the power of ice-core science.[6]

Getting stable isotope data from the Camp Century ice core took a lot of work and time. In CRREL's New Hampshire freezers, Langway and Dansgaard's team cut 1,600 samples of ice that spanned the entire depth (over 4,500 feet) of the core. They sampled the clear-ice parts of the core but steered clear of the basal ice—laden with varying concentrations of sand and silt—and the 11.5 feet of frozen soil.

In Copenhagen, Dansgaard and his team melted each sample of ice and analyzed the resulting water using his mass spectrometer. The team measured the ratio of the two most common oxygen isotopes, ^{16}O and ^{18}O. As Dansgaard had shown with his beer bottle experiment fifteen years before, and as Epstein showed with samples from the Site 2 core, that ratio faithfully mirrored the temperature of the snow as it fell onto the ice sheet.[7] A simple age/depth model, controlled by layer counting at the top of the core and based on Glen's flow law for the deeper ice, allowed them to estimate the age of each ice sample. Over several years, the team used the oxygen isotope data to build a proxy record of relative temperature through time for northwestern Greenland. Dansgaard was cautious and didn't try to calculate the actual magnitude of temperature change. There were too many uncertainties.[8]

On the basis of the similarity of oxygen isotope ratios in the core to those measured in pits dug into the surface snow near Camp Century, Dansgaard speculated that the bottom of the core contained ice remaining from the last interglacial warm interval. That was a time, more than 100,000 years ago, when both temperature and sea level were higher than they are today. In Europe, this interglacial is called the Eemian. It takes its name from the River Eem in Holland, where a sediment core taken in 1875 revealed unusual marine fossils in older sediment beneath the modern-day river muds. The shells in that older sediment belonged to mollusks that today live far south of the North Sea. Such warm-water fossils meant that the temperature of the water on the Netherlands coast must have been higher in the past. In climate records built from analyses of marine sediment, the Eemian goes by an alias, marine isotope stage (MIS) 5e.

Today we know that the Eemian interglacial occurred between 129,000 and 111,000 years ago, and that it was probably slightly warmer than today.[9] Coastal landscapes around the world preserve evidence of elevated MIS 5e sea levels, suggesting that ice sheets shrank as glacial ice melted. There are coral reefs now high and dry above the waves along the coast of Florida and the Caribbean.[10] On the island of Mallorca in the Mediterranean Sea, stalactites record seawater flooding caves at least seven feet above today's high tides.[11] That flooding began 126,000 years ago and lasted about 10,000 years. In semiarid southern Australia, deep beds of fossil shells dated to MIS 5e lie between seven and fifteen feet above the present sea level.[12]

Globally, sea level at the peak of the Eemian was as much as twenty-five feet above today's beaches; there are lots of data, but the measured sea level varies from site to site because of complicating effects, including the ups and downs of Earth's tectonic plates. Most experts think that six to twelve feet of sea-level rise during MIS 5e came from limited melting of Greenland's ice, but there are scarce data on which to base that conclusion. The rest resulted from the melting of the Antarctic Ice Sheet, mountain glaciers, and the expansion of seawater as it warmed.[13]

As Dansgaard discovered, Camp Century's ice faithfully preserved memories of long-vanished climates.[14] After the Eemian, the climate in Greenland cooled over tens of thousands of years. The greatest chill was between 25,000 and 15,000 years ago. Then a warmup began, but it was erratic. There were wild swings in isotope ratios, indicating that the temperature at which snow fell on Camp Century oscillated from warmer to colder and back to warmer again, ranging over many degrees but remaining, in general, below freezing. In Dansgaard's record, the last and most extreme swing began about 11,500 years ago, when the climate rapidly plunged back to nearly full-glacial cold. Ecosystems and glaciers in Europe and North America felt this chill and responded: ice advanced, and only cold-tolerant plants remained on the now-frigid landscapes.

This climate reversal, which lasted over a thousand years, is known as the Younger Dryas. Paleoclimatologists named this interval after a small, cold-tolerant Arctic tundra plant, *Dryas octopetala*, the leaves of which are common in pond sediments deposited around the North Atlantic region during this cold interval. Just before the Younger Dryas, rapid melting of the Northern Hemisphere ice sheets added massive volumes of freshwater to the North Atlantic. This freshwater, which is less dense than seawater, acted as a buoyant cap on the ocean, slowing its circulation. As a result, less warm tropical water moved northward. After more than a millennium, the chill broke, and for the next 3,000 years, the isotope record suggests steady warming with no cold snaps as the Pleistocene ice sheets finished thinning and retreating.

In the youngest 8,000 years of ice, there are comparatively minor changes in oxygen isotope ratios, implying climate stability on Greenland and, by analogy, on Earth. This interval is part of the Holocene (the last 11,700 years), during which agriculture arose and modern human civilizations developed and thrived. The uppermost 1,500 feet of the Camp Century core records the last 1,600 years of climate. First are the centuries of relative warmth that allowed the Norse to settle Greenland. Then, near the top of the core, oxygen isotope ratios reflect the chill of the Little Ice Age (after about 1200 CE), when canals in Europe froze and the Norse ultimately left Greenland.

Dansgaard's analyses of the Camp Century ice core, the first to penetrate any ice sheet, showed the power of water stable isotopes for documenting, in great detail, climate changes over at least the past 100,000 years. Camp Century's ice revolutionized our understanding of climate change over glacial-interglacial time scales. It was now clear that ice cores collected from the upper elevations of ice sheets could provide continuous, interpretable records of past climates.[15]

ICE AND SNOW cores preserve many of nature's antics. Snow isn't pure water, and neither is glacial ice. The wind carries, as small particles of dust, whatever we and the planet send into the atmosphere. A smattering of ash and sulfur from volcanic eruptions, salt blown in from distant oceans by storms, and radioactive fallout from atmospheric atomic weapons tests all mix with snow and eventually become entombed within ice sheets, as do lead from smelters and car exhausts, big organic molecules made only in chemical plants, and soot from slash-and-burn land clearing.[16] By counting seasonal layers in an ice core and modeling the flow of ice over time, scientists can date the ebb and flow of everything that has landed on an ice sheet. For example, we know when the Māori settled New Zealand because soot from their fires appears in Antarctic ice cores around 1325 CE.[17]

Camp Century's ice core provided a treasure trove of raw material for scientists to analyze—enough for them to publish nearly a hundred papers over seven decades. The topics of their studies varied widely. Some scientists looked to the ice core for data on both natural and human-induced fallout. Others considered the borehole left behind, using it to understand the temperature profile through the ice sheet and the speed and means by which the ice deformed and moved. Air content measurements down the length of the core suggested that the ice at Camp Century was more than a thousand feet thicker at the height of the last glaciation than it is today.[18]

The more than 4,500 feet of clear ice that makes up the bulk of the Camp Century ice core preserves every major volcanic paroxysm of the last 100,000 years. Acidity, sulfur dioxide, and tiny volcanic particles are the unmistakable fingerprints of ancient eruptions. Tiny bits of Mount Vesuvius, the eruption of which buried the city and citizens of Pompeii in the heat of an Italian summer day on August 24, 79 CE, clearly show up in glacial ice from 1,400 feet below the snowy surface of northern Greenland.[19] Volcanic ash flowed over Pompeii at a temperature of nearly 600°F.[20] Preserved in the ice core, today it's chilled to −24°F.[21]

In 1816, Indonesia's Mt. Tambora erupted so violently that bits of ash and sulfur-rich dust fell on Greenland, half a world away. That eruption gave the world the "Year without a Summer," during which snow fell in June, July, and August in New England, and people starved because they lost their crops to frosts, rain, and cold.[22] Similarly difficult and dismal weather covered much of Europe. Along a dreary Swiss lake in a summer filled with clouds and repeated bouts of chilly rain, Mary Shelley authored her dark novel *Frankenstein.* A century and a half later, Henrik Clausen, the Danish chemical engineer who had accompanied Dansgaard to Camp Century, found evidence of the Tambora eruption in eleven different Greenland ice cores, including the ice from Camp Century.[23]

Because of the way air masses circulate around the globe, Northern Hemisphere eruptions leave the largest and most easily detectable record in Greenland's ice cores. Analysis of the entire Camp Century core showed that the highest level of acidity occurred about the time of Julius Caesar's death (44 BCE), when by chance, the sky over Rome turned dark. Virgil, writing at that time, recalls that "when Caesar died, the Sun felt pity for Rome, as it covered its beaming face by darkness, and the impious generation feared an eternal night."[24] Scientists spent years speculating about which volcano(es) darkened Roman skies and emitted the acid found in Camp Century's ice. In the end, they concluded that a smaller eruption of nearby Mount Etna probably blocked the sun when Caesar passed. But it took four decades, six precisely dated and analyzed ice cores, and fieldwork on a remote Alaskan island for them to reach that conclusion.

In the meantime, scientists had found that multiple ice cores recorded two other closely spaced eruptions, neither of which happened in the year that Caesar died. The first of these eruptions (small, and perhaps from Iceland) occurred early in 45 BCE. Then, at the start of 43 BCE, a massive eruption poured acid-laced snow onto Greenland for two years. Measurements of sulfur isotopes in the ice showed that the volcanic emissions reached the stratosphere and circulated around Earth, which explained their long-lived effects. Scientists chemically analyzed minuscule bits of volcanic ash filtered from melted samples of ice dating to just when the large eruption began. These tiny shards were a perfect match with rocks from Okmok II, one of two volcanoes on a remote Aleutian island in the North Pacific—a place that now has, in addition to its volcanoes, an airstrip, a school, a post office, a power plant, and a population of thirty-nine people.[25]

There's strong evidence that the eruption of Okmok II triggered short-term changes in weather and climate that altered the course of history. Climate models considering the amount of acidity measured in the ice cores (specifically, sulfuric acid) suggest that two years of significant cooling and changes in rainfall followed the eruption. Other records of climate, such as the width of tree rings, indicate that the Mediterranean region and western North America were cold and wet in both 43 and 42 BCE. Writings from the time report that food shocks and famine soon followed, extending as far east and south as Egypt.

Soon after Okmok II erupted, the Roman Republic collapsed, and the Roman Empire took its place. Within little more than a decade, the Ptolemaic dynasty, which had long ruled Egypt, fell. Here was evidence that Greenland's ice preserved records of a climate event that "probably resulted in crop failures, famine, and disease, exacerbating social unrest and contributing to political realignments throughout the Mediterranean region at this critical juncture of Western civilization."[26] Linking the ice-core record with historical documentation reveals the power of a changing climate to alter the trajectory of human civilizations.

In 1980, data on carbon dioxide concentrations in icebound air bubbles began to emerge from ice-core studies. Twenty-one years after his Science Service interview, Bender's vision for using the composition of ancient air as a way to understand Earth's changing climate became a reality. The first data were noisy, but the pattern was clear: Ice Age carbon dioxide concentrations were far lower (about 50 percent lower, it seemed) than those of the present interglacial warm interval, the Holocene.[27]

Two years later, Hans Oeschger, a Swiss scientist and close colleague of Langway's, reported carbon dioxide concentrations stretching back 100,000 years, which his group had measured in ancient air bubbles locked in ice from the Camp Century core.[28] Their data revealed the composition of the atmosphere long before people had made any significant impact on planet Earth. The Camp Century record showed that Holocene carbon dioxide levels were consistent and high, at about 300 parts per million (ppm). Looking backward in time, concentrations begin to fall at about 10,000 years ago, bottoming out between 20,000 and 25,000 years ago—the last glacial maximum. In the few deeper and older samples, including two Eemian ice samples from just above bedrock, concentrations of carbon dioxide begin to rise again.

Within the resolution of the measurements at the time, the Swiss team showed that temperature and carbon dioxide concentrations in the atmosphere rose and fell together. Elevated levels of carbon dioxide correlated well with warm temperatures on Earth. Lower carbon dioxide levels coincided with cooler periods. The history of Earth's climate over the 100,000-year cycle that marks the pulse of glacial-interglacial dynamics was coming into focus as a result of rapidly improving analytical technologies and Camp Century's first-of-its-kind ice core.

The Swiss team's 1982 analysis of samples from nearly a mile of Camp Century ice core showed unambiguously that the concentration of carbon dioxide in the atmosphere today is higher than it's been at any time over the last 100,000 years. Antarctic ice cores pushed that limit back closer to a million years. Measurements from sources other than ice cores have taken the carbon dioxide record back tens of millions of years. Our atmosphere now has 50 percent more carbon dioxide than it did a century ago, and carbon dioxide levels are higher now than they have been since the mid-Pliocene warm interval (3–4 million years ago). Fossils found today on Ellesmere Island indicate that in the mid-Pliocene, giant prehistoric camels roamed a forested Arctic that was as much as 30°F warmer than it is today.[29]

If the past predicts the future—the mantra of geologists—then, given that carbon dioxide concentrations are rising dramatically in the atmosphere, climate warming should not be far behind. In the forty years since the first measurements of carbon dioxide concentrations in ice cores, that warming has become an unsettling reality.

WHAT ABOUT THE eleven and a half feet of frozen sub-ice sediment—the permacrete—from the bottom of the Camp Century ice core? Although unique, this material attracted far less scientific attention than the clear ice above it.[30] This lack of attention was puzzling to me for two reasons. First, geologists have described sediments for centuries, and in the 1950s and 1960s, the Army had conducted dozens of detailed studies of the permafrost around Thule and Tuto, including that within the permafrost tunnel. Second, Langway had received a $142,000 grant in 1969 with the title "The Analysis of the Deep Ice Cores and Sub-Ice Material."[31] Yet only four scientists used the frozen sediment at the bottom of the ice sheet for their research; three published their work.[32]

In 1980, Brian Whalley, a geographer in Belfast, Ireland, and an expert in glaciers and glacial processes, took a microscopic approach to studying the sub-ice sediment.[33] After Langway sent him envelopes of sand, Whalley used an electron microscope to characterize the shapes of grains at high magnification. His micrographs showed that many of the tiny sand grains from the top sections of the permacrete were angular and fractured, suggesting that ice had once transported them. But some other grains were smooth, the result of rounding from transport and abrasion. Whalley concluded that wind must have moved those sand grains across the landscape, and that the ice later swept them up as it advanced over Greenland. His paper was the first to hint that the ice was once gone at Camp Century—a groundbreaking conclusion that seems to have gone largely unnoticed at the time.[34]

In 1986, David Harwood, then a graduate student at the Ohio State University and now a professor of earth science at the University of Nebraska, received ice and sediment from the Camp Century core.[35] He processed it to extract the fossil shells of tiny photosynthetic algae called diatoms. Different species of diatoms live in fresh and salt water, and all make their shells of silica, which is quite hard and resistant to decay. Harwood found diatoms in one sample of the debris-laden basal ice and in two samples of the topmost sub-ice sediment (within twenty inches of the top of the permacrete). Most of the diatoms he found live only in freshwater. None of the diatom species he found were extinct—all live today in the Arctic.[36]

Harwood was looking small and thinking big. The existence of modern diatoms frozen in and beneath the basal ice was a key discovery. Sometime in the last few million years, freshwater must have flowed over the land surface now frozen beneath Camp Century. There's a statement at the end of Harwood's paper that should have caught a lot of people's attention because it implied that the Greenland Ice Sheet once shrank dramatically, then expanded and returned.

> These diatoms suggest the northwest corner of Greenland, and probably much more of the continent, was exposed during an episode of ice retreat associated with the Pleistocene interglacial period. In order to expose the Camp Century site during a Pleistocene interglacial . . . a reduction of at least one-third to one-half the present ice sheet volume must have occurred. This

> reduction implies that previous interglacials were warmer and/ or longer duration than the present Holocene "interglacial."[37]

Over nearly four decades, other scientists cited Harwood's paper only eighteen times. Few people seem to have noticed his idea that the Greenland Ice Sheet was dynamic and sensitive to changes in climate.

John Fountain, a young volcanologist at the University at Buffalo, The State University of New York, published the only paper describing the sub-ice sediment.[38] He briefly noted the layering, texture, and composition of the sub-ice material—well sorted and sandy at the top, icy with less sediment in the middle, and a heterogeneous, pebble-filled mass at the bottom. Fountain went on to characterize more than a dozen pebbles, each plucked from the outside of the permacrete core. They included granites (igneous rock), gneisses (metamorphic rock), and sandstones (sedimentary rock). Chemical analyses showed that the pebbles were similar to the rocks that crop out around the coast of western Greenland—high in silica, aluminum, and iron. His paper ended with a request for more drilling, stating that he needed samples of bedrock from beneath the ice sheet to put his findings from the core in a broader context.

It perplexed me that Fountain's paper hadn't appeared until 1981, fifteen years after the Army finished drilling the Camp Century core. Fountain and Langway both arrived at Buffalo in the mid-1970s: Langway in 1974, becoming chair of the Geology Department in 1975, and Fountain as an untenured junior faculty member. Fountain was the paper's first author; Langway was its last. The paper's authors (including Joe Wooden, who studied rocks from the moon for NASA) were clearly aware of this delay when they wrote in the paper, "The sub-ice material has received little geological attention as it represents samples of unknown lithologies and structures from beneath the ice. However, similar studies on grab samples from lunar rocks and meteorites have indicated that such investigations are not totally futile."[39] Still, I couldn't understand why it had taken so long to reveal what was in the sediment below the ice, so I called Fountain to ask him about his research. Over nearly an hour, I got an earful, not so much about the paper or the delay, but about Langway and the ice-core storage facility at Buffalo.

Fountain told me that Langway was insular, singularly focused on ice-core science, and could be impulsive. He was an administrator who wrote grants well, got money, made decisions, and ran people hard. Langway's research group, including the postdocs and graduate students who did the day-to-day work, kept to themselves and didn't interact with the rest of the Geology Department. During his thirty years at Buffalo, Fountain, like most of the other faculty, never set foot in the cold rooms where the ice cores resided. Although Fountain was the lead author of the Camp Century sub-ice paper, he'd never seen an ice or sub-ice sediment core. Langway handed him bags of pebbles to analyze. Wooden, the third author, who had dated the pebbles, told me in an email that he had no memory of the paper, the research, or the rocks.

Fountain shared recollections of Langway's resistance to "anyone new doing anything new" in ice-core science, of Langway "catching wind of a good idea and incorporating it into his next proposal," and of Langway saying out loud, "I own this field." Fountain clearly remembered that toward the end of Langway's career, things weren't going well in the Geology Department.[40] At the end of 1989, Langway resigned as chair of the Geology Department, a position he'd held for fifteen years, since he came to Buffalo.

KEEPING ICE CORES cold as they are moved from cold places to warm ones is a challenge. The Army had no problem when the cores were in Greenland since it could take advantage of nature and the nearly ubiquitous permafrost. Photos taken in the 1950s and 1960s show ice-core sections, each in its own aluminized tube about five feet long and five inches wide, stored on wooden racks in the permafrost tunnel at Tuto—a natural cold room. The Army flew the Site 2 core to the United States and stored it in SIPRE's walk-in freezers in Willamette, Illinois.[41] More cores soon arrived from Antarctica—from Byrd Station and Little America—in 1959.

In 1961, the ice and Langway, along with the rest of SIPRE, moved to the Army's newly established CRREL facility in Hanover, New Hampshire. By then Bader was no longer an Army employee. He'd taken a research professorship at the University of Miami, although a grant from the Army continued to pay his salary.[42] With Bader's departure, Langway assumed an increasingly prominent role in managing the growing archive of ice cores and in parceling out samples to others who made specialized measurements. He also spent time in Greenland with the drill crew at Camp Century.

Soon, tubes of ice had begun arriving from Camp Century. Drillers filled more than 100 tubes in 1961 as the first hole penetrated 600 feet into the ice sheet. The second hole generated 150 more tubes of ice, and the third hole, the one that penetrated the entire ice sheet, filled over 1,000 tubes. When the last core segment arrived from Camp Century, nearly a mile of ice, and more than eleven feet of frozen soil from below the ice, filled the freezers at CRREL. More than 7,000 feet of ice from a second Byrd Station ice core, recovered by Ueda's team between 1967 and 1969, also arrived in Hanover. In the early 1970s, more short cores from Greenland followed: Dye-3 (1971), Milcent (1973), and Crete (1974). The original CRREL freezer could hold only about 400 tubes. To keep all the rest of the ice securely frozen, CRREL rented a large commercial freezer in Littleton, New Hampshire, sixty miles north of its headquarters.[43]

In 1959, the Army drafted Bill Vest, who had just completed a geology degree. The military assigned him and a group of other science-trained soldiers to Fort Belvoir, Virginia. Vest's first several-months-long deployment took him to Greenland, and he kept going back. For three summers (1960–1962), he worked with Langway and the drilling team. When Vest left the military for graduate

Camp Century ice cores in aluminized tubes moving off the ice sheet and down the ramp to Camp Tuto on May 11, 1965. The cores traveled across the ice sheet in a heavy swing. *U.S. Army photograph taken by Specialist Fifth Class, Charles T. Bodey and provided by Jon Fresch.*

training in geology, Langway hired him to log ice cores for two more summers. When we talked, Vest told me that "Chet Langway was just the greatest guy to work with you could ever imagine." But he also said that Langway "marched to a different drummer than other government employees. . . . Well, he was an independent cuss, if the government or the Army said we are going to do this he would fight it and that's why he didn't stay at CRREL."[44]

When the University at Buffalo hired Langway to chair its Geology Department in 1975, he took the ice-core collection with him—by then about four miles of ice. At the same time, the U.S. National Science Foundation began funding a new ice-core storage facility at Buffalo, directed by Langway. The first funds arrived in

April 1975. Automatic grant renewals followed every few years until 1991.[45] Over sixteen years, the foundation gave Langway, and the University at Buffalo, more than a million dollars to operate the Central Ice Core Storage Facility and Information Exchange. By 1989, the facility held six miles of ice.[46] Langway housed some of the ice on campus in a walk-in freezer, but most sat in a commercial freezer downtown.[47]

Erick Chiang curated the core repository from 1975 until 1979. He wrote of Langway, "He was one of the few glaciologists at the time who had the foresight to create storage facilities to preserve ice cores obtained from Greenland and Antarctica. . . . Langway was one in a loose confederation of international scientists who understood the value of ice cores and the stories that they could tell not only of climate change . . . but also of other impacts of human development over time."[48] An email from Guy Guthridge, a former program manager at the National Science Foundation, tells very much the same story while echoing Fountain's assessment of Langway: "Chet basically was the savior of ice cores when few were interested. NSF once was not forthcoming with funds for the curation work. Chet called us to say, 'It's easy to dispose of ice cores. You just put them out on the loading dock.' NSF paid up."

Chiang told me that "Langway and his colleagues are true scientific pioneers whose stories would tell of the foundations on which the realities of climate change are built." Yet, despite the science he and his collaborators did and published, it's not clear that Langway believed humans capable of significantly changing climate. A reporter profiling the Buffalo laboratory in 1988 provides four paragraphs of evidence suggesting the reality of human-driven global climate change, but concludes, "Langway, however, is emphatic about not drawing conclusions for the future from current ice core research. 'You can speculate like hell, but we just haven't done enough research to have all the answers.' "[49]

For more than thirty-five years, while he was working at SIPRE, CRREL, and the University at Buffalo, Langway controlled access to almost every ice core from Greenland and Antarctica, the majority of which Americans had drilled. As a condition of funding, the National Science Foundation mandated that ice in the Buffalo facility be available to anyone with a valid need. In reports about the facility, Langway wrote,

> Samples of the ice cores are made available to any interested scientist that is funded through OPP [The U.S. National Science Foundation's Office of Polar Programs] for this purpose or has an on-going research effort that would benefit from obtaining polar ice core samples. To obtain samples, in the first case, submit a proposal through OPP. In the latter case, write to the author.[50]

But, as Fountain noted, and as I found in speaking to polar scientists who trained and worked between 1970 and 1990, for some people, getting ice and information from Langway and his core repositories wasn't as simple as these public statements suggested.

Lonnie Thompson, now a world-renowned ice-core scientist at the Ohio State University, told me that when he was a graduate student in the 1970s, Langway refused to give him samples of ice from the Camp Century ice core.[51] Thompson got the ice he needed only after Jay Zwally, a program manager at the National Science Foundation from 1972 to 1974, intervened.[52] That ice, and Langway's recalcitrance, shaped Thompson's career.[53]

Using samples from the Camp Century and Byrd Station cores, Thompson examined tiny windblown dust particles embedded in their glacial ice.[54] These particles were far smaller than the width of a human hair. What Thompson found changed our understanding of how landscapes, some far from the ice, respond to glaciation. In both ice cores, the particle record matched the water stable oxygen isotope record perfectly: cold glacial periods were dusty times on Earth; warm interglacials were not. Soon after Thompson completed his measurements, he encountered Langway at a small workshop of ice-core scientists. When Thompson told Langway about the dust results, Langway advised him not to present his findings, claiming they would end Thompson's career. Thompson gave the talk anyway.[55]

Although Thompson eventually got Camp Century ice for his doctoral research, Langway told him not to expect any more from the next generation of Greenland ice cores. Thompson solved the core-access problem by avoiding Langway and the Buffalo core repository entirely. Instead of analyzing ice from Greenland or Antarctica, Thompson built his career studying high-elevation ice caps (some over 18,000 feet above sea level) in the tropics.[56] When Dansgaard learned of what Thompson wanted to do, he wished him luck, but thought his ideas would never work and that he and his team might die trying to recover these cores. But Thompson soon sent ice from the heights of the Andes to Copenhagen, where Dansgaard measured the oxygen isotopes. Their data showed that ice from the tropics preserved a faithful record of ancient climates in equatorial regions where no other similar data exist.[57]

By 1989, it had been fifteen years since almost every American ice core, and Langway—described at the time by a *Los Angeles Times* reporter as a "chain-smoking geology professor who has a mountain named after him in Antarctica"—moved to Buffalo.[58] During those years, the NSF had focused on growing the ice-core analysis community in the United States, an effort consistent with its mission to promote American science with American tax dollars.[59] Yet it still seemed that Langway, when distributing samples, was favoring senior scientists while often refusing requests from American graduate students. Large amounts of ice went to labs, many in Europe, where most of Langway's closest collaborators worked. Langway particularly valued Dansgaard's stable isotope facility, Oeschger's ability to measure the composition of ancient air, and Clausen's geochemical prowess in identifying past volcanic eruptions, all critical for synchronizing records between ice cores.

A 1989 newspaper article, one of the last about Langway, identifies him as "one of three 'patriarchs' in deep ice core studies—along with Willi Dansgaard of Denmark and Hans Oeschger of Switzerland." Langway claims that "collectively, the three of us could monopolize all of these kinds of studies." The reporter's analysis differs: "Now, however, Langway finds himself in competition with hundreds of researchers seeking a slice of the National Science Foundation grant budget. The field has opened up, drawing interest from scores of related science disciplines."[60]

In 1990, the National Science Foundation stopped automatically renewing Langway's grant to manage the national ice-core storage facility. Instead, it issued an open call for proposals from the entire U.S. scientific community. Such re-competition is the norm for large, expensive national facilities and usually happens every five to ten years. For the ice-core storage facility, it was long overdue.

Julie Palais is an ice-core scientist and a woman with a unique view of the American ice-core community. As a graduate student, she went to Buffalo and tried, but failed, to get Byrd Station ice from Langway for her doctoral thesis (which she completed in 1985). After her graduate work, Palais spent five years at the Ohio State University examining ice cores for evidence of volcanic eruptions.[61] Then she moved to the National Science Foundation, where she spent the next twenty-six years managing both Arctic and Antarctic research programs. As one of her first assignments at the foundation, Palais oversaw the ice-core storage facility re-competition.

The foundation received only two proposals. Langway—by then a thirty-four-years-plus veteran of ice-core science—sent one from Buffalo. The other came from Mark Meier, Tad Pfeffer, and Walter Dean. The first two were glaciologists at the University of Colorado, so they knew ice well, but neither had a track record of ice-core analysis. Dean was a geologist with the U.S. Geological Survey who had long collected and archived lake and marine sediment cores. The foundation had to decide: should it try something new, or stick with what it had?

Later that year, the National Science Foundation made the decision to fund the Colorado proposal, and by doing so, changed the course of American ice-core science dramatically.[62] Thirty years later, when I spoke to Palais, she recalled, in a single word, why the repository went to Colorado: access. Both the proposal reviewers and the panel of experts brought together to make the final decision were concerned that, just as people who had worked with him indicated, Langway played favorites, which meant that getting ice for research depended on who you were as much as the science you wanted to do.[63] Because the directors of the new American ice-core facility studied glaciers, not ice cores, the National Science Foundation and the ice-core community hoped that the new management team would help level the playing field.

It took three years after the re-competition to open the new U.S. National Ice Core Laboratory in Lakewood, Colorado, a collaborative effort between the University of Colorado, the National Science Foundation, and the U.S. Geological Survey. The foundation built the facility from scratch and paid to move the ice-core collection across half a continent—a necessary one-time investment. Thousands of tubes filled with dozens of ice cores found their way from New York to the new facility in Colorado, including the entire cores from the first and second holes at Camp Century. But the lowest 131 meters of the third and final Camp Century core, including ice, basal ice, and the sub-ice sediment, never arrived. Nor did the bottom forty-three meters of the Byrd Station core, the first to reach the bottom of the Antarctic Ice Sheet. This was a crucial loss to ice-core science. Although the missing samples represented less than 10 percent of the Camp Century core, they also represented about 80 percent of the climate record because the annual layers, squeezed by the ice above, become thinner and thinner with depth.

When the National Science Foundation selected Colorado, rather than Buffalo, to oversee the American ice-core collection, an aging Langway lost control of the archive he'd built over decades. In the 1990s, the next generation of American polar scientists was collecting new cores from both Greenland and Antarctica—GISP2 and Taylor Dome—and Langway didn't participate in those projects. Much of the ice from those cores was going to newly built labs in the United States for processing and analysis. The field of ice-core science that Bader envisioned, and that Langway had built, was quickly leaving him behind.

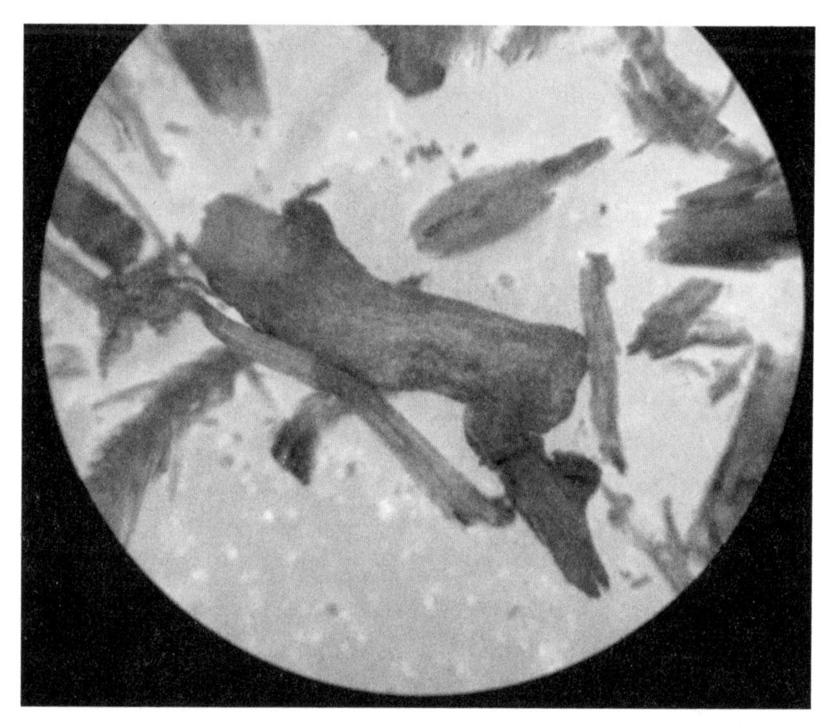

Looking into the microscope on July 26, 2019, we saw plant parts from the upper sample of the Camp Century sub-ice sediment core. *Photograph by Andrew Christ.*

WHEN GREENLAND WAS GREEN

We use ice as a history book, that's what fascinates me.

—DORTHE DAHL-JENSEN

On a warm September afternoon in 2017, Dorthe Dahl-Jensen sat down next to me on a picnic bench in Buffalo, New York. She's a faculty member at the Niels Bohr Institute in Copenhagen and one of the world's leading ice-core researchers. She was also one of the first women in what, since its inception, had been the male-dominated field of polar exploration, ice coring, and ice-core analysis.[1] For two days, we'd been sitting inside listening to talks at a U.S. National Science Foundation–sponsored workshop titled "How Stable is the Greenland Ice Sheet?" Dahl-Jensen turned to me and said, "Paul, I know you are interested in basal ice and debris from the bottom of the Greenland Ice Sheet. We have a lot of that in Copenhagen. You should come and have a look." I was intrigued, and planned to see the material and Dorthe when I next went to Denmark.

A year later, when I headed to Knutz's workshop in Denmark, I sent Dahl-Jensen an email. She was in Greenland and couldn't respond. Another year passed, and then an email from her appeared, titled simply "Workshop on Oldest Ice." It was an invitation to join her and about twenty others in Copenhagen in April 2019. The list of invitees included six American, one Chinese, and thirteen European scientists. Near the bottom of the email, Dahl-Jensen stated, "We have ice from Dye-3, Camp Century, GRIP, NGRIP and NEEM." Missing was any mention of the Camp Century sub-ice sediment.

The workshop met early on a spring morning at the Niels Bohr Institute in a small conference room. Talks about methods and results of previous work on basal ice filled the first day. Most of the presentations reviewed published science that I knew well—but not Dahl-Jensen's. She presented detailed results that I'd never seen before: DNA data from the Camp Century basal ice suggesting that molecules from tundra shrubs and trees had found their way into that material. If the data were right, then the ice that now covers northern Greenland had once given way to a boreal ecosystem teeming with life. Dahl-Jensen also presented a series of dates from the silty basal ice that suggested a wide range of possible ages ranging over the last million years. This was as exciting as it was confusing.[2]

The next morning, the ten of us who remained piled into a large van and headed to the suburbs of Copenhagen, going against the morning traffic. After perhaps a thirty-minute ride, we arrived at the Danes' new ice-core storage facility—a nondescript, one-story metal building in a gated industrial park. It was a brand-new, state-of-the-art facility, purpose built for storing ice cores. The refrigerant, they told us with pride, was carbon dioxide.

The ice-core freezer was a building inside the building. Donning winter hats and gloves, which seemed odd on a warm April morning, we went inside. The freezer was the size of a small supermarket. There were hundreds of foam-lined cardboard boxes sitting on wooden pallets and covering most of the floor. Each box contained sections of ice cores. Written on the side of each box was a name, a date, and a series of numbers. Pushed against the side of the room were row after row of empty steel racks, each four shelves high. The place was so new that the curators had not yet placed the boxes of ice on the shelves. A separate room with saws for cutting ice was slightly warmer, but still well below freezing.

Stored here were decades' worth of ice cores from around the world. Enough ice, someone told us, that if you laid all the tubes end to end, they'd make a line more than twelve miles long. We walked around, dumbstruck by the sheer volume of ice. I took a photograph on my phone as Dahl-Jensen held a piece of basal ice core above her head; the dark silt bands were clearly visible against the white freezer walls. It was −17°F and I was uncomfortably cold. Dahl-Jensen put the core down, and her husband, Jørgen Peder Steffensen, led a few of us to the far corner of the freezer. Steffensen is a tall, bearded man with a deep laugh, a physics degree, and a love of ice. If there's a reigning king of Greenland ice coring and ice cores, it's Steffensen. He oversees the ice-core storage facility in Copenhagen and travels to Greenland several times a year.

In the corner, on pallets, sat two foam boxes, encased in old and tattered cardboard held together by black webbing straps. Stacked inside and towering over the open top of each box were a dozen or so clear glass jars, each about the size of a person's head, and each capped by a white plastic lid. Handwritten white labels adorned the outside of each jar. Written in neat black cursive, they seemed at the time like a code we didn't know how to crack: 1062–1, 1063–4, and 1061-D5. Inside each jar was a thick plastic bag containing a cylindrical lump of brown material. Each jar also held a small paper label slipped inside it that read "October 12, 1972"—probably the day when someone at CRREL cut these samples and placed them in the jars. Large frost crystals obscured the details of the samples inside. Steffensen turned to us and announced calmly, "These are the Camp Century sub-ice samples."

I don't remember saying anything, but I do remember picking up one of the samples. It was in a bag outside of its cookie jar. I stared at the brown lump intently, gently feeling the roughness through my thin gloves and the thick plastic. In the cold, I slowly felt my stomach drop. I was holding in my hands both human and

I first saw the Camp Century sub-ice samples in Copenhagen on April 4, 2019. Joerg Schaefer is to my left. Obscured is Nicolaj Krog Larsen.

natural history. Standing there shivering, both from the cold and from anticipation, I knew this was a once-in-a-lifetime experience and opportunity. In my forty years as a geologist, this was the only set of samples that ever sent shivers up my spine.

There is something otherworldly about material from the bottom of an ice sheet, material collected more than a half century ago, material that most people had years ago given up on ever finding. Looking at these samples, you wouldn't know why—they were brown, sandy, and badly freezer-burned bits of soil, any one of which you could hold in the palm of your hand. Vanished both from people's minds and from ice-core storage facilities, which usually know exactly what samples they hold and how to find them, these cylinders of frozen soil had now almost magically reappeared. Eleven and a half feet of sediment from the bottom of the ice sheet had been hiding in plain sight, safe in Copenhagen in a pair of old, brittle brown cardboard boxes labeled "Camp Century sub-ice."

That afternoon, our group (rapidly shrinking as people ran to catch trains and planes to head home) talked about what to do next. We agreed that test samples were in order. We'd start with the bottom and the top of the sub-ice sediment core. The Danes would cut us small pieces of these two parts of the core, keeping the rest as an archive. They would ship those pieces to Vermont, in an insulated box lined with cold packs designed to keep ice cores frozen for several days. Once they arrived in our lab, we'd melt most of each sample, recover the water and sediment for analysis, and send parts of each onward to a dozen other labs in the United States and Europe. As spring turned to summer, we waited for the frozen soil to arrive.

I'D BEEN INTERESTED in glaciers since my first year at Williams College in 1980. Several years later, my senior thesis focused on the demise of the Laurentide Ice Sheet in northwestern Massachusetts. In a rusty gray 1968 Volvo, I went out to survey gravel pits, excavations for new homes, and eroded streambanks that revealed deposits of sand, gravel, and clay—all left by the melting ice sheet as it exposed a barren and eroding western Massachusetts landscape 15,000 years ago. I dug into glacial till deposited at the bottom of the ice and imagined silt-clouded ice-marginal lakes, dammed by lobes of the ice sheet that by then filled only the deepest mountain valleys. Through these observations, I re-created in my mind a glacier and a landscape that no longer existed.

For the next forty years, I continued to study glaciers, ice sheets, and Earth's changing climate while expanding my research into human-landscape-climate interactions. In the 1990s, as a graduate student, I became interested in the field of cosmogenic isotope geochemistry. I was one of the first group of geologists to venture into that field because it provided a means to determine when, and at what rate, events happened on Earth's surface. Cosmogenic isotopes opened the door to dating the comings and goings of glaciers and ice sheets, in some cases extending back in time to the dawn of the Pleistocene, 2.7 million years ago.

Cosmogenic isotopes are exceedingly rare forms of elements, produced as energetic cosmic rays from outer space interact with atoms in both Earth and its atmosphere. For decades, such isotopes were mainly the purview of nuclear physicists. In the early 1990s, geologists began working with other scientists who were measuring cosmogenic isotopes in rocks and soil with the hope of answering long-standing and confounding questions in earth science. It didn't take long for our community to realize that these isotopes—primarily beryllium-10 (^{10}Be) and aluminum-26 (^{26}Al)—were robust tools for determining the timing and pace of landscape changes at and near Earth's surface.[3]

Isolating ^{26}Al and ^{10}Be from rock and soil is time consuming, hazardous, and requires a specialized laboratory.[4] This process takes weeks to months. In most cases, after grinding samples, we use sieves, magnets, and days of soaking in warm, weak acid to remove every other mineral except quartz (the shiny, hard, clear material that makes up beach sand in most places). We then dissolve the purified quartz for several days in hot hydrofluoric acid, the only acid that attacks this robust mineral. In our clean rooms—wonderful for allergy sufferers because there is no pollen or dust—we carefully purify the resulting liquid using centrifuges to remove solids and ion-exchange resins to separate beryllium and aluminum from impurities such as titanium, iron, calcium, sodium, and magnesium.

The results, after weeks of work, are small, dried pellets about the size of a flea. Each pellet sits at the bottom of its own test tube until we heat it, driving off any remaining water. We grind the pellets into powder and pack each sample, using a hammer and tamper, into a small hole at the top of a stainless-steel cylinder about the size and shape of a .22-caliber rifle round. Our lab work is then complete.

The analysis of our samples requires atom counters—repurposed multimillion-dollar particle accelerators. Once favored instruments for nuclear physics, today they don't have sufficient power to answer many interesting physics questions. Nevertheless, with the proper modifications, they make excellent mass spectrometers, able to distinguish exceptionally rare isotopes from their more common siblings. The accelerator mass spectrometers on which we measure samples can find one atom of ^{10}Be in a million billion atoms of ^{9}Be. That's one part in 10^{15}. Or, in human terms, it's like finding one person in more than 125,000 million Earths, each as populous as ours is today. These hefty instruments require warehouse-sized buildings and a team of engineers to keep them running.

The application of cosmogenic isotopes to dating glacial deposits can be simple or complex. Determining the ages of boulders left behind by melting glacial ice is often straightforward: measure the concentration of isotopes in a sample and divide by the production rate of each isotope (something that is well known and changes predictably depending on altitude and latitude). Dating rock or soil first exposed on the surface, then buried by ice, and then exposed again is more complicated. Using radioactive cosmogenic isotopes that decay away at known but different rates, it's possible to calculate both the minimum duration of initial exposure and the minimum time of burial.[5] The differing decay rates, or half-lives, of ^{10}Be and ^{26}Al (1.4 and 0.7 million years, respectively) make the pair particularly useful for dating exposure followed by burial on glacial time frames, hundreds of thousands to a few million years.

My first experience dating boulders involved those left behind by debris flows near Lone Pine, California—debris flows that originated from glaciers extending toward the semiarid valley bottom from the highlands of the Sierra Nevada.[6] Through my research, I was able to calculate that glaciers in the area had reached their maximum size between 21,000 and 26,000 years ago—data that match well with estimates for when glaciers around the world, including Greenland's ice sheet, reached their maximum. Consequently, this was also a time when global sea level was at its lowest, more than 400 feet below today's shoreline.

My interest in the Arctic started in 1995 with a project in Baffin Island, Canada, during which my colleagues and I (led by Thom Davis, a glacial geologist and lake-coring expert, and including then–master's student Kim Marsella) discovered what geologist (and our lab manager) Lee Corbett termed "ghost glaciers"—ice that once occupied the uplands of Baffin Island.[7] These cold-based glaciers had barely left a trace, just a boulder here and there; yet the combination of ^{26}Al and ^{10}Be isotope data showed us that thick glacial ice had covered much of the island's landscape for most of the past million years.

A decade later, I was talking to Richard Alley, a Penn State glaciologist, about what we learned on Baffin Island and how to apply that knowledge of isotopes and ghost glaciers to the Greenland Ice Sheet. Drilling holes to the bed of the ice was difficult, time-consuming, and expensive, and at the time, neither of us knew the whereabouts of the Camp Century sub-ice material. But then we thought, nature was doing the sampling for us as the ice sheet eroded its bed. That eroded sediment made up the till that ended up in and below ice melting out around the ice-sheet margins and choked the meltwater streams pouring from the edge of the ice sheet.

To test our ideas, I wrote a proposal with Tom Neumann, at the time a University of Vermont professor, who had experience in Greenland fieldwork. The reviewers were skeptical but intrigued, and the National Science Foundation funded what they termed our high-risk, high-reward idea. After several weeks of fieldwork, we came back from Greenland with nearly a hundred fist-sized cobbles and then, a few years later, dozens of bags of sand from rivers draining the ice sheet. The isotopic results (there were very few cosmogenic nuclides in most of our samples) told us that in many places, the ice sheet was warm-based and thus deeply eroding its bed.[8] But a few of the cobbles had much higher isotope concentrations, and had ratios of ^{26}Al and ^{10}Be indicative of burial, just like the sediment that once lay beneath the ghost glaciers of Baffin Island.[9] The isotopes were starting to reveal complex behavior at the bed of Greenland's ice sheet.

Then we tried another approach, our first foray into ice cores. In 2010, my colleagues and I measured the concentration of ^{10}Be in silty ice from the bottom of another Greenland ice core—taken in 1993 at the GISP2 core site at the summit of the ice sheet.[10] Our data showed that there, where the ice was two miles thick, the concentration of ^{10}Be on sediment entrained by the ice was extremely high. Thus, we concluded there was little erosion. The ice sheet must have been present for much of the Pleistocene to protect the soil, and must have remained mostly cold and frozen to the rock and sediment below.

By then, I was getting more and more interested in how the Greenland Ice Sheet worked and how it had behaved in the past. Before I worked on the GISP2 basal ice, I'd never thought much about Camp Century and its ice core. I had read Dansgaard's 1969 stable oxygen isotope paper, but knew little about the sub-ice sediment or the basal silty ice. I started asking around, and everyone that I spoke to had long ago concluded that the sub-ice sediment was gone. As a relative newcomer to Greenland, who was I to question what seemed like common knowledge? Our lab went back to finding ghost glaciers, this time in northern Greenland near Thule.[11]

On July 23, 2019, a yellow-and-red DHL van pulled up in front of Delahanty Hall, a slate-clad brick campus building in Burlington, Vermont, home to our laboratory, which includes several clean rooms. The driver brought a white foam cooler to the third floor. It was a hot and humid summer day in Vermont, and we'd been waiting for this shipment since late April.

Drew Christ, then a doctoral student, and I cut the strapping tape and pried off the lid of the cooler. Inside were pink long-lasting cold packs and three plastic bags of still-frozen sediment. Pulling out the bags one by one, we glanced at their contents. One held nothing more than a pile of frosty sand. The other two contained half cylinders, each a few inches wide and a couple of inches long. One half cylinder was from the top of the sub-ice core. The other was from the bottom of the core, and the frosty sand was the freezer burn shed from the top sample while it was in storage—a replicate sample of sorts, as it came from the outside of the core.

We stopped, opened the last bag to see the core material, and quickly resealed it. The stench of diesel fuel and solvent was overwhelming. More than a half century later, the smell of the 1966 drilling fluid persisted.[12] We quickly placed all three samples in a large white chest freezer. Catching our breath, we gathered some tools and supplies.

We removed one sample at a time from the freezer, carefully photographed it alongside a ruler for scale, and then weighed it. Working from a detailed plan we had created since we first saw the sub-ice samples in April, we began to take each sample apart slowly and methodically. Our goal was ambitious: to decipher the history of northwestern Greenland's ice sheet and the land beneath it from a few thousand grains of sand, a couple of tablespoons of mud, and less than an ounce of water, which, when frozen, filled the pores between the grains of sediment. Our experience the year before with Knutz's far less precious marine core material from Baffin Bay told us that our approach was likely to work, in large part because geoscientists had an analytical toolbox in 2019 that no scientist in 1966, even in their wildest dreams, could ever have imagined.[13]

To avoid breathing the irritating and toxic fumes of diesel and TCE, we did most of our work in fully exhausting fume hoods like those you'd find in a chemistry lab. First, we pried a few small frozen chunks, about the size of a dime, from the samples and put them right back in the freezer. We planned to embed some of the chunks in epoxy, polish them, and then examine the grains using

The upper sandy material, filled with plant fossils, was the first Camp Century sample we processed, July 23, 2019.

an electron microscope that magnified the material over 1,000 times. We saved other chunks for luminescence dating, which would tell us the last time the material saw the sun. The remaining chunks we left overnight in a walk-in cooler at 40°F to melt, each sealed in its own thick plastic bag.

The next day, we extracted the water from the now-melted permacrete. First, we cut a small hole in the bottom corner of each plastic bag. Then we drained the water and sediment into a clean centrifuge bottle. We spun the bottles at high speed for ten minutes, hoping that the sediment would settle and the water would come to the top. It worked, but only partially. With an assortment of pipettes and a syringe, we removed as much of the meltwater as we could, leaving behind only that liquid that filled the spaces between the soil grains. What little water was left was thick with silt. We transferred that water to clean test tubes and set them back in the cooler for later chemical and isotopic analysis. When we finished, we placed the sediment samples in a just-warm oven so that we could find out how much the sediment weighed when dry.

Once the sediment had dried, we began our routine process of preparing it for analysis. Christ picked up the first sample and a Munsell color chart—a book of what looks at first glance like thousands of tiny paint chips systematically organized by color—which is a standard tool for soil scientists. Flipping through the pages with a spatula full of dried sediment next to the chart, he found the best match and read the color codes to Leah Williamson, an undergraduate student who worked with us for several years. The first sample matched up with 2.5 YR 4/2, *weak red*, a color indicative of weathered soil and the presence of oxidized iron—rust.

For analysis, we needed to separate the sediment grains that made up each sample by size: pebbles you can pick up one by one, sand grains that pour through your fingers, and powdery silt and clay. I grabbed a clean black plastic oil tray and a brass sieve nine inches across. Grains smaller than two millimeters (about one-twelfth of an inch) would fall through the sieve's metal lattice; anything larger would stay in the sieve. Williamson dumped the sample into the oil tray, and we added water and stirred. Then, one of us poured the thick brown mixture into the sieve as the other held it over another tray to catch all the rinse water. We washed the sample through several different sieves with more water, and finally transferred the sediment using clean water into five labeled plastic bins—one for each grain size from the sample: gravel; coarse, medium, and fine sands; and silt with clay.

We left the bins with the samples on the lab cart for a moment, washed and put away the tools we didn't need, and got out a clean bucket to catch the rinse water from the final sieving. This last rinse held all the silt and clay, the finest material that had passed through the last sieve—one with openings we could barely see, perhaps twice the width of an average human hair. We worked to save every grain of these samples because, in total, we had only a pound of material to share between more than a dozen labs.

We were ready to start sieving again when I stared absentmindedly into one of the plastic bins. In the few moments we'd been cleaning, the sand had settled out, and the water was almost clear. I looked again, and then once more. There were small black bits, longer than they were wide, floating on the surface of the water. All I could see at that moment was a room in the old Perkins Building where I had processed lake cores with my students in the 1990s and early 2000s.[14] Back then, we were trying to determine both when the Laurentide Ice Sheet had left Vermont and the frequency of major storms through the Holocene by dating plant parts preserved in lake sediment cores.[15] One small twig, leaf, or needle held enough carbon for a reliable measurement of carbon-14, the isotope used for carbon dating. The trick was finding the fossils in the meters of fine black lake sediment that filled our core tubes.

Usually, we would slowly scan the split lake cores and find a twig here and a leaf there in the otherwise homogeneous, sticky, and foul-smelling pond muck. But sometimes nature didn't cooperate, and we found none of these precious fossils. When that happened, we used a trick that Thom Davis, an old hand at lake sediment core analysis and the person who introduced me to Arctic science on Baffin Island, showed me: take a scoop of sediment, put it in a beaker of clean water, stir thoroughly, and wait. First the sand settles to the bottom, then the silt, and, more slowly, the clay. Plant parts, being less dense than the water, float to the top—a birch twig, an alder leaf, a spruce needle. These fossils were what we needed to get a reliable radiocarbon date.

Staring at the bin of sediment from Camp Century, I mumbled something out loud about plant parts and organic material. That earned me a puzzled gaze from both my colleagues, so I grabbed a disposable plastic pipette and a small white tray. After a few moments of capturing the floating black bits, I handed Christ a sample of the material to look at under the microscope. He stared down the eyepieces of the binocular scope for perhaps thirty seconds and said nothing, then looked up and blurted out a string of expletives.

What he saw confirmed my gut feeling. These first two test samples from the Camp Century sub-ice sediment were full of plant parts: bits and pieces of tundra vegetation, twigs from small willow trees and berry bushes, mosses, leaves, and fungal spores. Fifty-three years and almost a month after the core came out of the ice, we had found the frozen remains of an ecosystem that somehow, no one else who had looked at the sediment had ever noticed. It told us unambiguously that the ice in northwestern Greenland had vanished sometime in the past and then returned—exactly the conclusion that those looking at the sediment in the 1980s had come to from other lines of evidence, but which the community had largely ignored.

It was an astonishing and completely unanticipated finding that made for an unforgettable afternoon. The Camp Century drillers had gotten exceptionally lucky. No other ice core since has recovered so much soil or so many plants. In a few moments, our chance discovery of these plant pieces forever changed what we knew about the stability of Greenland's ice sheet.

THE EXPLOSION OF technology, and the specialized expertise needed to create and interpret data today, make it impossible for one person to know and do it all scientifically. When there is limited sample material (and that's the case for cores of both ice and sediment), teamwork is the norm, and samples are allocated to maximize the science and minimize the amount of material used. The goal is always the same: have some material left for another day, when someone can ask a different question or use a new analytical approach.

We were lucky. The Camp Century sediment core was virtually untouched—few people had sampled it, and thanks to Langway and the Danes, especially Dahl-Jensen and Steffensen, almost all the sub-ice sediment had remained safe and frozen since 1966.[16] To put this sub-ice material in perspective, consider that in 1969, Americans began landing on the moon. In just a few years, they ferried 842 pounds of moon rocks to Earth, an important part of the science mission of the Apollo program.[17] Today, there are 69 pounds of Greenland sub ice material.[18] This stuff is, as the credit card folks put it, priceless.[19]

Within a few days of sieving the samples and finding the plant parts, we began sending subsamples to an international, collaborative team, many of whom had attended Dahl-Jensen's Oldest Ice workshop—a conglomerate of geologists, physicists, geochemists, and geochronologists. We were following, without knowing it at the time, Bader's directive to diversify the study of ice cores—more than twenty-five years after his death.[20] In contrast to those who drilled the Camp Century core in the 1960s and analyzed the sediment and ice in the following decades, our team was nearly half women.

Among our group, Eric Steig, a professor at the University of Washington, measured the stable isotopes of oxygen and hydrogen in water from the core, along with carbon and nitrogen isotopes in the organic material. These data revealed the temperature and elevation at which precipitation fell and the types of plants that once lived on the landscape nearly a mile below Camp Century. Elizabeth Thomas, a professor at the University at Buffalo (in the same geology department where Langway worked for years), used ultrasensitive gas chromatographs teamed with mass spectrometers to identify traces of waxes that used to cover the plant leaves. Leaf waxes are like fingerprints; they reveal what types of plants grew in the fossil ecosystem now frozen under the ice.

Dorothy Peteet is a paleobotanist at Lamont-Doherty Earth Observatory. By comparing specimens in her reference collection with those we isolated from the two test samples, she and Ole Bennicke (from GEUS in Denmark) determined many of the plant species that once covered the landscape below Camp Century. Nico Perdrial, a geochemist, uses electron microscopes and X-rays to study the texture and size of sediment grains and to map the distribution of elements and minerals in samples of rock and soil—critical information for understanding the exposure and weathering history of sediment.

Tammy Rittenour is a geology professor who directs the Utah State University Luminescence Laboratory. Luminescence dating liberates electrons trapped by defects in the atomic structures of minerals to reveal when sunlight last illuminated the samples. It's useful back to at least a million years, but mandates that Tammy work mostly in the glow of dim red lights preparing samples for dating. She provided dates telling us when the ice re-covered the land beneath Camp Century.

Paul Knutz and Tonny B. Thomsen, both geologists, work at GEUS using lasers to sample mineral grains and mass spectrometers to date minerals, producing data that are critical for tracing where in Greenland sand grains in the sub-ice sediment came from. Alan Hidy, a geologist, and Marc Caffee, a physicist, use massive particle accelerators teamed with mass spectrometers at Lawrence Livermore National Laboratory and Purdue University to measure isotope ratios of cosmogenic aluminum and beryllium and thus date the burial of sediment by ice. Pierre Henri Blard measures isotopes of neon and also studies the shapes of grains much as Whalley did forty years earlier. Jean-Louis Tison worked with Blard to log the core pieces in the freezer, painstakingly sketching their layering and measuring their density to estimate ice content.

Lee Corbett manages the cosmogenic nuclide lab that I first built in 1993. Our clean chemistry lab is a place of test tubes, Teflon, and exceptionally clean air, filtered four different times before it washes over the samples we process. It's also home to lots of strong acids. Lab work here means always wearing gloves and safety goggles. Stylish, bright yellow full-body smocks, along with closed-toed rubber Birkenstocks, complete the outfit and keep everyone safe. Our job is to do the weeks of chemistry needed before samples are ready for ^{10}Be and ^{26}Al measurements on Hidy and Caffee's accelerator mass spectrometers. Joerg Schaefer, a geochemist at Lamont-Doherty Earth Observatory, does similar work; together, we replicated analyses of the Camp Century samples to ensure the data were reliable.

Every one of us, and our labs, had the same goal: to understand what the sub-ice sediment could tell us about the history of northwestern Greenland and the ice that covers it today. Together, we asked a fundamental set of questions: Had the Greenland Ice Sheet melted before, and if so, when? And what was the ecosystem like when sunlight instead of ice covered northwestern Greenland? No one set of analyses could answer these questions, but there is power in numbers. The more we knew about the sediment, the more detailed and more accurate the story we could tell about Greenland and its ice.

Finding the plant fossils certainly quickened our pace. Not wanting to lose the momentum of the April meeting in Copenhagen, we scheduled an international workshop in Vermont for mid-October, barely ten weeks after the test samples arrived. That's an awfully fast turnaround for labs that typically take months, even years, to process samples and generate data. The workshop would use data from the two test samples to refine our analytical approach and secure funding to analyze the remainder of the sub-ice sediment. Everyone was working overtime to arrive at the workshop with data.

On a warm October evening in 2019, forty-two scientists from around the world came together in a renovated warehouse near the Burlington waterfront alongside Lake Champlain. They filled the large, open room with soaring ceilings and twelve-inch wooden beams. Many were eating strong local cheese and drinking beers named New Math, Loud Places, and Dare Tonight as we kicked off a two-day meeting focused entirely on Camp Century and its sub-ice sediment core.

The next morning, the science started. Eighteen of those attending brought new data, generated from the samples we distributed in late July. Careful examination of the twenty-eight pieces of sub-ice sediment core now in Copenhagen had refined, but hadn't changed, the description that Fountain had published in 1981.[21] The top of the core (the first test sample) was sandy and layered, most likely deposited by moving water in the absence of ice, consistent with Harwood's diatoms.[22] There was an ice-rich, low-density layer in the middle. The bottom (the second test sample) was stiff, compact glacial till.

The science that came from a pound of frozen sediment was stunning. Over the course of the day, people talked about leaf wax analyses, mineral compositions, uranium-lead ages, and argon dates. There were physical descriptions of the sub-ice sediment and models of ice flow to complement organic carbon content determinations, rare isotope measurements, luminescence data, and water stable isotope ratios.[23] Many of the data were only days old. By evening, all the new data were public. Thirteen hours of science in one day—more focused attention than the Camp Century sub-ice sediment had ever received.

Both the upper and lower test samples contained plant fossils. The upper sample contained more and better-preserved plant pieces than the lower sample. Every species of plant we could identify in the sub-ice sediment still lives today in the Arctic tundra ecosystem that rings the ice-free margin of Greenland. The leaf waxes, while more degraded in the older, lower sample, also suggested the presence of tundra vegetation. Water isotopes indicated that rain and snow fell at warmer temperatures, and likely at far lower elevations than the current elevation of the ice sheet today, which made sense. The ice was gone when the water seeped into the soils. Then, the soil and water froze when the ice returned.

The upper and lower samples also showed dramatically different degrees of mineral weathering from exposure to liquid water near the surface. The scanning electron microscope images and X-ray maps of elements coating soil grains in the lower sample—the till—showed that a mixture of clay, iron, and aluminum covered most sediment grains—an observation indicative of a long-lasting, stable soil exposed near Earth's surface. The pore water we extracted from the till contained high concentrations of dissolved minerals, suggesting that the water had spent a long time in contact with the sediment.

In contrast, sand grains from the upper sample looked fresh under the electron microscope. The grains had few coatings, and the pore water between them contained far less material in solution than the water in the till. Together, these data suggested a much shorter residence time for water in the upper than in the lower sample. This observation implied that the upper material had not spent much time sitting on the landscape without ice over it. Although the upper and lower samples looked different, dating and analysis of minerals in the two samples gave the same ages. suggesting that they shared the same source rocks. The tiny grains of hornblende, a dark aluminum- and silica-rich mineral, in both samples crystallized between 1.8 and 2 billion years ago. Indeed, we began to suspect that the upper sediment was the result of a stream eroding into the till when the ice was gone. As the sand bounced its way downstream, the grain coatings had eroded away.

Paired measurements of ^{10}Be and ^{26}Al showed that the ice sheet deposited the till (the lower sample) no more than 3 million years ago. Luminescence data showed that sunlight hadn't shone on the mineral grains in that till for at least the past 1.7 million years. Together, these ages bracketed the timing of an ancient glaciation to sometime between 3 and 1.7 million years ago—we now knew when ice deposited the till at the bottom of the core. Perhaps that till was evidence of the first ice sheet to cover northwestern Greenland. The upper, sandy part of the core was younger. In October, there were only cosmogenic data and detailed descriptions of the samples, but they clearly showed that moving water deposited the uppermost material less than a million years ago. How much less we didn't yet know, except that radiocarbon analysis told us that the plant fossils, and thus the upper sediment, were older than 40,000 years, the practical upper limit of that dating method with small samples.[24]

The history and story of the sub-ice sediment were taking shape fifty-three years, three months, and a few weeks after coring ended at Camp Century. We now knew that the permacrete preserved a history of Greenland's ice sheet that stretched back as much as thirty times longer than the 100,000 years preserved in the overlying ice. The frozen soil preserved not only evidence of when the ice covered the land, but also evidence of when it did not. Fountain, Whalley, and Harwood had come to the same conclusion when they studied the core in the 1980s, but they had no idea when the ice had vanished. Now, we had dates and plant parts and the implications were significant.

Sometime in the last million years, the ice at Camp Century melted without people pumping carbon dioxide into the atmosphere. Nature alone created a planet warm enough for long enough that tundra plants had replaced ice in northwestern Greenland. If Greenland's ice could melt without people, and with the atmospheric carbon dioxide concentration at 280 ppm, then the ice sheet should be increasingly vulnerable as atmospheric carbon dioxide concentration continues its steady and rapid climb above 400 ppm, more than 30 percent higher than before the Industrial Revolution. Greenland's ice vanished at least once before, and thus, it could vanish again—especially now that Earth's climate is warming so much and so quickly.

During the second day of the workshop, working groups of scientists, self-organized by expertise and interest, created a detailed plan for analyzing the rest of the sub-ice core by sampling the remaining twenty-six sections still stored in Copenhagen. All the new data showed clearly that the sub-ice sediment held the secrets of a changing climate over millions of years—stretching back in time far beyond the extraordinary record preserved in Camp Century's ice. On the basis of this thinking and the preliminary data, we spent more than a year crafting, and then submitting, a nearly 200-page proposal to the U.S. National Science Foundation to sample, analyze, and understand those remaining parts of the core.

That evening, after the workshop was over, three dozen people filled our home in Burlington for dinner. Steffensen, who five months before had showed me the Camp Century sub-ice sediment in the Copenhagen freezer, sat down at the end of our crowded dining room table. I sat down opposite him, hoping to chat. Our university's environmental reporter, Josh Brown, placed a tape recorder between us.[25] For forty-three minutes, Steffensen told the story of how he remembered the Camp Century and Byrd Station cores getting to Denmark. Now gray haired, he was a graduate student when the ice arrived.

Steffensen started by saying that in 1994, he believed that few people asked for samples from Camp Century anymore. The new GRIP and GISP cores, both deeper and at the center of the Greenland Ice Sheet, were all the rage. He remembers hearing from Henrik Clausen, his doctoral advisor, about an unhappy phone call from Langway. Clausen was an ice-core chemistry expert, a professor at the University of Copenhagen, and Langway's lifelong friend; their relationship had developed in the 1960s after Clausen cut thousands of samples of ice from the Camp Century core in CRREL's cold rooms.[26] Steffensen told us that "[Langway] was so fed up . . . that he picked up the phone and, it's sort of anecdotal, but I was told by my senior, Henrik Clausen, Chet put it this way: He said, Hank, you come over here and get some ice because otherwise it goes into the lake. We are so sick and tired of it."

At that point, Steffensen recalls, the samples started coming to Denmark. The first shipment to Copenhagen included twenty crates of ice. Steffensen remembers two of the crates. Someone had oddly labeled them "Camp Century, sub-ice." The second shipment that year consisted of two, twenty-foot-long containers, both filled with ice. Between 1994 and 1996, Langway continued to send more of the Buffalo ice-core collection to Denmark.[27] The containers held some of the Camp Century and Byrd Station ice cores, including all of the lowest—and thus oldest—sections. Now, most of the Camp Century climate record, and all of the sub-ice sediment, was in Copenhagen. There's no doubt these samples came from Buffalo. Forty years later, a box labeled "sub-ice sediment" still retains its original labels from the State University of New York at Buffalo and Langway's ice-core laboratory there.[28]

Langway knew ice sheets and how they worked. He was keenly aware that the deepest ice and the sediment below were treasure troves of information about climate and ice-sheet history. The National Science Foundation had requested, and expected, that Langway would send the entire American ice-core archive to Colorado from Buffalo. But he didn't. His refusal to transfer the most informative and difficult-to-replicate sections of the Camp Century and Byrd Station cores to Colorado was an act of loyalty to his European collaborators that assured them of continued unfettered access to those samples at a time when the American ice-core community was growing, changing and diversifying rapidly.[29] For more than two decades, no one, at least in the United States, realized that the Camp Century sub-ice sediment core still existed.[30] Unbeknown to the polar science community, the two boxes of Camp Century permacrete remained safely frozen, unopened, and, it seems, forgotten somewhere in Copenhagen.

In 2018, the Danes were preparing to open their new centralized ice-core storage facility. For decades, they had stored their collection of ice-core samples in a variety of freezers scattered around Copenhagen. Now that all the ice was in one place, Steffensen and his team took inventory. His dinner plate now empty, Steffensen turned, looked at Josh, and said,

> So, then I go through the whole record and say: what is this Camp Century "sub-ice"? So, I went down, and I had a look into the boxes, and I saw these jars. . . . Well, when you see a lot of

cookie jars, you think who the hell put this in here? No, I didn't know what to make of it. But once we got it out, we picked it up to see these dirty lumps, and I said: what is this now? And all of a sudden it dawned on us: Oh shit, this is sediment underneath it. The "sub-ice" is because it's BELOW the ice. Whoa.

Seventeen months after the workshop, our team's first paper, titled "A Multimillion-Year-Old Record of Greenland Vegetation and Glacial History Preserved in Sediment beneath 1.4 km of Ice at Camp Century," appeared in the *Proceedings of the U.S. National Academy of Sciences*.[31] There were eighteen authors, representing thirteen institutions and four countries. The take-home message was simple: Greenland's ice sheet under Camp Century had vanished at least once in the past million years, and when it did, tundra covered the landscape. The ice had come, gone, and come back again in northwestern Greenland.

A few months after the paper came out, I received an email from the program officer in charge of the Arctic Science program at the National Science Foundation. The foundation was funding our proposal to analyze samples throughout the entire sub-ice sediment core from Camp Century. Our team had support for several more years of research, including funding for seven faculty and a half-dozen graduate students, to better understand the Camp Century permacrete. In the fall of 2021, Christ traveled to Copenhagen and spent a month in the Danish ice-core facility with colleagues from Belgium, France, and Denmark, slowly and methodically cutting samples—all under red safelight and all kept well below freezing. He flew back to Vermont with pieces of each of the twenty-six samples; nearly forty pounds of still-frozen soil and ice returned to the United States for the first time in nearly three decades.[32]

Soon after, Cat Collins, one of three Vermont master's students working on the project, began imaging each sample using a CT scanner in a CRREL freezer. Those X-ray images preserved a detailed digital record of the layering in each of the samples before we melted them. I like to think of Collins's work as a homecoming for the permacrete, which had left CRREL for Buffalo more than forty years before. After she completed the scans, we melted each frozen core segment, extracted the water, sieved the sediment, and sent aliquots to labs around the world.

Juliana Souza, another Vermont master's student, filtered the once-frozen water from each sample. By measuring the dissolved elements, she found that the upper sediment, the intermediate ice layer, and the lower sediment had quite different chemistries: the ice was pure, probably of glacial origin, while the chemistry of the lower-sediment pore water indicated that it had remained in contact with the soil for far longer than the water from the upper sediment, consistent with the older age of the lower sediment and the results from the test samples. Halley Mastro, the master's student with expertise in plants, separated the fossils from each of the samples and painstakingly identified as many as she could, working with Peteet in the United States and Ole Bennike, one of the world's experts in Greenland fossil plants, in Denmark. This time, they found more species of plants and several different insects, including the midges that swarm around your head in Greenland's short summer.

In our lab, we purified quartz from every sample and spent months processing it so that Hidy and Caffee could measure the ratios of ^{10}Be and ^{26}Al. A depth profile through the till showed no indication of surface exposure in the absence of the ice sheet. Most of the upper sediment had for several thousand years been a stable surface exposed to cosmic rays, perhaps a small terrace alongside a stream. The upper foot of sand was isotopically homogeneous, probably stirred by the seasonal freezing and thawing that characterize Arctic permafrost soils on which tundra vegetation grows today.

In 2023, while almost three dozen people (faculty, graduate students, technicians, staff scientists, and undergraduates) were working on samples, and others, including filmmakers and podcasters, were telling the story, the most consequential single data point of the project so far appeared. Rittenour had measured the latest exposure age of the uppermost sample using her luminescence readers in Utah. The last time this sample had seen the sun was about 416,000 years ago (with an uncertainty of 38,000 years).[33] Now we knew when the ice was gone from beneath Camp Century, information that had eluded us, and others, for more than a half century.

Rittenour's age lands right in the middle of MIS 11, one of the longest, but not the warmest, of the interglacial warm intervals.[34] Atmospheric carbon dioxide concentrations were high then (about 280 ppm), but not extraordinarily so. In contrast, sea level rose higher during MIS 11 than during any other recent interglacial, as did the oxygen isotope ratios of seawater. Both of these findings are irrefutable evidence of shrunken ice sheets.

Benjamin Keisling, a geoscientist at the University of Texas, used all this information from the Camp Century sub-ice sediment to constrain a coupled climate and ice-sheet model, which he ran dozens of times with different, but completely plausible, parameters. His results showed that when the ice over Camp Century vanished, sea level rose at least five feet, and perhaps as much as twenty feet, as meltwater from Greenland poured into the ocean. The data were compelling: ice had left northwestern Greenland when the climate was similar to today's.

Langway and Clausen were right. The lower parts of the Camp Century core had been worth saving. The Danes' expertise at preserving this unique material, and their willingness to repatriate samples nearly thirty years later, allowed paleoclimate science to bloom more than a half century after the Electrodrill hit the bottom of Greenland's ice sheet below Camp Century. In 1969, nearly a mile of ice core from Camp Century enabled Dansgaard and Langway to parse the past hundred thousand years of climate change. Now, eleven and a half feet of sub-ice sediment have extended that timeline back millions of years. The frozen soil revealed that ice from below Camp Century vanished at least once in the past, long before people found and began to burn oil, gas, and coal.

As witness to Greenland's diminished ice sheet, the permacrete from the bottom of the Camp Century ice core provides a quiet but stern warning. Greenland's ice is fragile, and there was a time, as the plant fragments in the sub-ice sediment so plainly show, when Greenland was, unambiguously, green. Unless we change our ways, the island will be green again.

Aerial view of a dying ice cap in the highlands of western Greenland. All of the past winter's snow was gone by July 12, 2008. It's grown too warm now for a glacier like this to survive—only sediment-laden ice remains. This glacier no longer has an accumulation zone.

ICELESS

> As a species, we've somehow survived large and small ice ages, genetic bottlenecks, plagues, world wars and all manner of natural disasters, but I sometimes wonder if we'll survive our own ingenuity.
>
> —DIANE ACKERMAN, *New York Times*

With climate change and planetary warming come all sorts of what the military euphemistically calls collateral damage. For planet Earth and those of us living on it, that damage often comes quickly and noticeably, driven by extreme events: wilting droughts, overwhelming floods, searing heat domes, frigid Arctic blasts, devastating hurricanes, uncontrollable wildfires, and unprecedented marine heat waves in which ocean-surface temperatures top 100°F. In many cases, the extra energy provided by our warming climate makes once-ordinary weather extraordinarily damaging—destroying property, devastating communities, and killing people. Climate change is an accelerant, making common what in the past was rare.

Collateral damage also sneaks up on us, coming slowly, almost imperceptibly, as sea level around the world rises fractions of an inch each year—in lockstep with glacier melt and a warming planet. Compared with extreme weather disasters, sea-level rise gets less attention in the press because it doesn't lend itself to dramatic videos or hair-raising stories; rather, the daily beat of the tides and the movement of sand along the shore mask the gradual rise well—until a hurricane or nor'easter strikes. Then, the subtle and creeping effects of what seemed like minor changes in water level quickly become both noticeable and disastrous. An extra foot of water can be the difference between life and death—flooding the only road off an island, knocking a home off its foundation, or pouring into New York City's subway tunnels.[1]

Since the beginning of the Industrial Revolution, sea level has risen a bit less than a foot as Earth has warmed. As the rate of warming increases, sea-level rise will accelerate, too, as ice on land melts and as ocean water, which has soaked up 90 percent of Earth's climate change–induced excess heat, continues to expand.[2] That's a problem, because by 2100, when a child born today has retired, sea level could be as much as several feet higher. The millions of people living close to the coasts will find their homes, villages, and fields under salt water. Today, thermal expansion of warming ocean water alone accounts for between one-third and one-half of sea-level rise—a clear downside to the effective oceanic heat sink.[3]

Vanishing ice drives, in one way or another, many climate-related impacts on our lives. The line is simplest to draw between melting ice on land and the increasing volume of water in the ocean because it's direct—as in a bathtub: when water goes in, the level rises. When ocean volume increases, the water flows onto coastal land. Just under 10 percent of Earth's population lives within thirty feet of sea level.[4] Keeping ice sheets frozen matters to people living thousands of miles away from the polar regions.

As Greenland's ice melts, trouble, in the form of rising waters, will come to every coast. Greenland's ice holds enough water to raise global sea level more than twenty-four feet, and that, it turns out, is a problem especially for low-lying coastal megacities such as Jakarta, Kolkata, Osaka, Tokyo, Hanoi, and Shanghai.[5] The top twenty vulnerable cities (ranked by population exposed to sea-level rise) are all in Asia and at least 8,000 miles from southern Greenland, but that distance doesn't matter. If Greenland alone loses most of its ice, the resulting sea-level rise will put a half-million square miles of land underwater and displace several hundred million people.[6]

More subtle, but just as important, is the role of ice and snow as a fundamental part of Earth's thermal regulation. Their bright white surfaces reflect incoming solar energy and thus cool the planet. Wet snow is darker than dry snow. So, as the surface of the ice sheet becomes increasingly saturated by meltwater in the summer, Greenland's ice reflects less sunlight and absorbs more, accelerating the melt. Once the ice along the edges melts away, dark rock replaces reflective ice, and the warming increases further. At these places, where the ice is newly gone, it's as though someone turned up the local and regional thermostat. Eventually, the collective sum of that Arctic warming has global implications.

The loss of ice and snow at high latitudes results in Arctic amplification, an alliteration describing the increased warming of the Arctic compared with the mid-latitudes. Although the Arctic has often been thought to be warming about twice as fast as the global average based on models and older data sets, a recent analysis of near-surface temperature records suggests that it's warming at least four times faster today.[7] Once Arctic amplification sets in, the melting of snow, sea ice, and ice sheets accelerates in a vicious cycle.

We've started that cycle already. Greenland's ice sheet is shrinking, and the sea ice that once covered the Arctic Ocean has, over the last several decades, gotten markedly thinner, younger, and less extensive.[8] The town of Upernavik in northwestern Greenland once had only a sea-ice runway that was useful for much of the year. When the sea-ice season became shorter and the ice less predictable, they blew the tops off the island's two mountain peaks, used the rock to fill the valley between, and built a year-round airport above their town with a runway so short there's no room for error.

Arctic warming also matters to people living at lower latitudes. In Europe and North America, winter snows now come later in the fall and melt earlier in the spring. The time ice covers lakes has similarly shortened, and some lakes have stopped freezing altogether.[9] Greenland's melt influences ocean circulation, reducing heat transport from the equator to the poles. Today, one of the only parts of the world that is cooling is the North Atlantic south of Greenland—probably as a result of diminished warm ocean currents flowing northward. Loss of high-latitude snow and ice cover also affects the behavior of the jet stream, a band of strong wind that blows at 30,000 feet above Earth's surface. The jet controls the movement of air masses around the globe and thus weather thousands of miles away from the Arctic.

As scientists looking for useful analogs of our warming future, we had limited knowledge of how much Greenland's ice sheet melted during prior intervals of interglacial warmth, the most recent of which started about 130,000 years ago. There are some clues from forest DNA found in the Dye-3 ice core from South Greenland, the isotope geochemistry of marine sediment off southern Greenland, and cosmogenic isotopes in the rock beneath the GISP2 ice-coring site in the center of Greenland.[10] But none of these cores had dating firm enough for us to know exactly when Greenland's ice sheet was smaller than today.

Without a solid age, it's impossible to reliably link evidence of ice-sheet change to the climate record. This ambiguity left the scientific community largely dependent on combined models of ice sheets and climate to envision the speed and distance of glacial retreat when the climate warmed in the past. There were several generations of ice-sheet models, but they didn't agree well on what it took to melt Greenland's ice. The models also disagreed about when, where, and how much ice on the island survived prior warmings. That is, until we knew the age of the frozen ecosystem preserved in the Camp Century sub-ice sediment.

Now, with Rittenour's luminescence dating, the cosmogenic nuclide data, and Keisling's model results, we know when, for how long, and at least how much of Greenland's ice melted during MIS 11's warmth. The soil in the sub-ice sediment hadn't seen the sun since 416,000 years ago—a time when, as the plant parts tell us unambiguously, the land under Camp Century was free of ice. The cosmogenic data indicate that the sediment containing the plant parts remained exposed for no more than 15,000 years after the ice went away. Keisling's model of the ice sheet mandates that losing the ice from Camp Century would have raised sea level at least five feet. Even on the low end, that's enough water to flood 20 percent of Charleston, South Carolina, and 80 percent of New Orleans, as well as all the low-lying areas of Rio De Janeiro. But the model doesn't stop there—equally plausible solutions indicate that almost the entire Greenland Ice Sheet could have melted away at the same time Camp Century was free of ice, raising sea level more than twenty feet.

For geologists, the past is our way to predict the future. And Camp Century's sub-ice sediment gives us a reality check. The warmth and melting that took the ice out from under Camp Century during the beginning of MIS 11 probably looked a lot like the climate conditions that are shrinking the Greenland Ice Sheet today. Then, as now, the cold, dry center of the ice sheet continued gaining mass for a while as the lower, warmer edges shrank substantially, especially those on the west side, which has the drier and sunnier parts of the island.[11] Slowly, the melt season grew longer as the climate warmed. Over centuries, the portion of the ice sheet that lost mass every year expanded. Ice-albedo feedback (Arctic amplification) drove much of this warming and ice loss. Eventually, the nearly mile-thick ice under Camp Century melted away, and the first tundra plant emerged, perhaps from a seed carried by a warm wind.

Glaciologists track the mass balance of ice sheets by estimating the weight of snow that falls, the amount of ice that's lost around the edges, and the volume of the ice. When soldiers filled Camp Century in the 1960s, glaciologists had only the most rudimentary data—a few transects across the ice sheet with snow pits and snow density measurements. These data allowed an approximation of how much mass the ice sheet gained from snowfall every year. But there was no reliable way to measure ice loss by calving or the volume of water pouring off the ice sheet in the hundreds of meltwater rivers draining Greenland. Seismic lines provided a few estimates of ice thickness, allowing a rough calculation of total ice volume. But every value needed was no more than a crude estimate.

But the lack of data didn't stop Bader from trying to determine the mass balance of the Greenland Ice Sheet. Using every piece of data he could find (much of it from SIPRE's work throughout the 1950s), he concluded, in 1961, that Greenland was probably gaining mass, but admitted that his answer was quite uncertain—because at the time, he didn't have enough data, nor did he have observations taken over a long enough time.[12] NASA, working with German colleagues, eventually got the data Bader was missing. In 2002, the agency began the Gravity Recovery and Climate Experiment, or GRACE. They launched a pair of thousand-pound satellites into orbit and used them to measure changes in Earth's gravity from space. The GRACE instruments were so sensitive that they easily detected the removal of water pumped from the ground during California's extended drought.[13]

For a decade and a half, the satellites repeatedly flew over Greenland, beaming back gravity data and thereby measuring the mass of the ice sheet. As the ice sheet shrank, so did the gravitational force it exerted on the satellites. They could "feel" the snow accumulating in the center of the ice sheet and disappearing around the edges. Processing the data revealed areas of rapidly flowing ice around the margins emptying into deep fiords. There, massive ice cliffs calved giant icebergs that slowly drifted toward the open ocean as they melted.

The GRACE data aren't encouraging. The Greenland Ice Sheet's mass balance is changing—it's gone negative as we warm the Arctic. Now Greenland is losing a lot of ice, both by melting and by calving icebergs. From 2002 to 2016, Greenland lost an average 280 gigatons of ice per year (a gigaton is a billion tons).[14] Although it's hard to visualize such abstractly large numbers, NASA does its best to put this ice loss in perspective: "Every year Greenland sheds 10,000 fully loaded U.S. aircraft carriers of ice. . . . In the 15 years GRACE was watching Greenland, enough ice was lost to cover Texas in a sheet of ice 26 feet high."[15] According to another estimate, as the climate warmed between 1992 and 2018, the Greenland Ice Sheet lost 3,900 billion tons of ice.[16] That's enough meltwater to raise sea level around the world by approximately half an inch, or to fill an Olympic swimming pool for one in every five people on the planet.[17] None of this is good news for coastal residents. Every drop of water that once sat on Greenland and now enters the ocean raises global sea level.

As ice melts away, the bedrock on which it sat responds to the lightening load and rises. Fifty-four high-precision GPS (Global Positioning System) stations ring Greenland. Solid steel bolts secure their antennas to massive, solid bedrock outcrops. Each GPS station is a radio receiver that continuously collects data from satellites orbiting overhead. With those data, the stations calculate their positions over and over. Their GPS units are sensitive enough to measure the ups and downs of Earth's crust to small fractions of an inch. The GPS network easily measured Greenland's rise during the warm summer of 2010, during which anomalously large amounts of melting happened.

The daily averages show a distinct annual cycle. Snow in the winter increases the weight of the ice, and the land below sinks. As the snow melts in the summer and the load decreases, the land rises. If you map the GPS stations and study their data over time, the pattern is clear. Southern Greenland is rising faster than the rest of the island, as that's where much of the melting is occurring. Only recently has ice loss begun in northwestern Greenland.[18] Camp Tuto is long gone, but the road to the ice front, and the ramp of gravel leading to the ice, still exist. However, both the road and the ramp are now in pieces because much of the ice below them has melted as the climate has warmed.[19]

In the late 1990s, Volkswagen commissioned a similar ramp nearly 800 miles south at a site it called "660" where the ice met the land.[20] The company planned to build a snow and ice track on the ice sheet to test new cars in winter conditions, away from the prying eyes of competitors. But, after improving the bone-jarring gravel road from Kangerlussuaq to 660 and building a ramp up to the ice, Volkswagen never tested cars there. The company gave up on the project in the face of a warming climate and irrefutable evidence that the ice at 660 was rapidly melting. In just twenty years, the ice margin has retreated back so far that there's now a ramp to nowhere. Near 660, a Google street view shows isolated sections of the road at the tops of nearly vertical ice cliffs formed as the sediment-covered glacial ice below melts away. A small bulldozer is working on the ramp, doing its best to mend the failing road. The ice sheet itself is an increasingly long hike from where you can safely park.

This rapid ablation appears likely to end Kangerlussuaq's tourist business. Visitors fly from Copenhagen to Greenland on vibrant red Air Greenland planes, landing on the American-built World War II airstrip. Then, many board four-wheel-drive buses for a five-hour round trip to catch a glimpse of the ice sheet. They might even walk onto the ice, if the ramp and road allow, before their return to the settlement for a musk ox barbeque. Such tourism may not see many more years as ice-sheet melting accelerates.

It's not just the ice at 660 that's melting quickly; so is ice miles inland from the margin and the now-decapitated ramp. Some of the water from that melting pours into crevasses and moulins, ending up in and under the ice sheet. From there, it flows through natural icebound tunnels before emerging in roaring torrents at the edge of the ice—turbid with silt and frigid at just above the freezing point of water. The water is so cold that my hands went numb within seconds when collecting samples of sediment from these outwash streams. Other meltwater flows over the ice surface, cutting sinuous channels across the ice. Some of those channels end in icebound lakes, which have a habit of emptying quickly and without warning through hidden cavities in the ice below. The water and the ice are all an otherworldly shade of blue, turning toward gray when the sun approaches the horizon late in the evening.

Some of the meltwater enters the tributaries of Akuliarusiarsuup Kuua (the Watson River), eventually ending up in Kangerlussuaq, which is about twenty miles away. The road from 660 largely follows the same valley as the water. To the south of the town, a double span bridge connects Kangerlussuaq to roads on the other side of the river as well as to a lake for swimming, a gravel pit, and a dump. During the summer of 2012, a persistent high-pressure system lingered over Greenland, bringing continuous sunshine to daylight that lasted twenty-four hours. As a result, the temperature soared to over 70°F around the ice-sheet margin. When the wind was blowing enough to keep the bugs at bay, I even did fieldwork in shorts that year. During the warmest few days, satellite data suggested that 97 percent of the ice sheet's surface melted.[21] There was liquid water in the snow even at Summit Station, at the peak of the ice sheet, almost two miles above sea level. Torrents of meltwater flowed into Akuliarusiarsuup Kuua and, on July 12, 2012, swept away the bridge and a bulldozer working to save it. Built in the 1950s, the bridge had survived every prior melt season.

Ice cores from the summit of Greenland vouch for the past rarity of such widespread melting. Events like the one that happened in 2012 form distinctly solid, bubble-free, clear layers that stand out in ice cores. There are data on melt layers from the GISP2 core extending back 10,000 years.[22] During the warm middle Holocene, between about 8,000 and 4,000 years ago, Earth's changing orbit allowed the sun to deliver more solar energy to Greenland than it does today. Warm spells then caused the uppermost few inches of snow at Summit to melt, on average, once every 50–75 years. In the last few thousand years, as the climate cooled, melt events became much less frequent, averaging only one every few hundred years at Summit.

As Greenland rapidly warms in step with the rest of the Arctic, melt events are becoming increasingly more frequent. In 2019, the snow again melted at Summit. In 2021, it rained, astonishing the scientists there, who posted photographs online of the camp's windows covered in raindrops. In 2022, temperatures again rose above freezing at the top of the ice sheet, and the snow began to melt. Four melt events at Summit in a decade had no precedent in the last 10,000 years.

Of course, the 2024 concentration of carbon dioxide in Earth's atmosphere, over 420 ppm, is equally unprecedented. The last time the atmosphere held such a high concentration of carbon dioxide was in the Pliocene, more than 3.3 million years ago.[23] Sea level then was at least twenty feet higher than today, and for much of the Pliocene, Greenland had little, if any, ice. Since we now know from analysis of the Camp Century sub-ice sediment that nature alone melted a mile of ice at a time when there was only 280 ppm carbon dioxide in the atmosphere, it's hard to see Greenland's ice sheet surviving in anything like its present extent as carbon dioxide levels rise several parts per million each and every year.

TODAY, CLIMATE SCIENTISTS are swimming in data, with satellites mapping, day in and day out, the shrinking extent of ice sheets and sea ice to determine Earth's albedo (how much incoming sunlight reflects from the planet), and thus the warming potential. Other satellites monitor deforestation, which changes both albedo and carbon uptake from the atmosphere by plants. Automated floats monitor surface-water temperature and acidity in the world's oceans. Sensors determine concentrations of greenhouse gases, such as methane and carbon dioxide, in the atmosphere. Extensive, vetted global temperature records are now available online at the touch of a button. Month by month and year by year, the data are piling up.

Using these global data, the World Meteorological Organization showed that the eight warmest years (going back more than a century) were 2015 to 2022.[24] 2023 made that a nine-year streak. On July 3–6, 2023, Earth had its four hottest days, shattering prior records. Twenty-one of the thirty hottest days ever recorded occurred during July 2023. Paulo Ceppi, a climate scientist at the Grantham Institute in London, believes that July 4, 2023, was probably the warmest day Earth has seen in the last 125,000 years—that is, since the peak of the last interglacial period, MIS 5e, when sea level was at least ten feet above the current level.[25]

This heat followed the warmest June ever recorded, with unprecedented wildfires burning across eastern Canada and an ocean heat wave warming surface waters across the North Atlantic to levels unseen in our 170-year-long record of ocean temperatures. Phoenix, Arizona, had thirty-one consecutive days with high temperatures over 110°F—another record. Plastic playground slides there registered more than 160°F—sufficiently hot to burn a child's skin in seconds.[26] Waters around the British Isles were almost 9°F above normal in June, while water in a coastal lagoon in Florida topped 100°F. All this heat is also melting ice. Antarctic sea-ice cover in 2023 was at record low levels, while the United Nations reported "off the charts" melting of mountain glaciers.

Scientists have been warning about the impacts of shrinking glaciers around the world for decades now. In a 1973 letter to the *Journal of Glaciology*, Bader remembered his time in South America and, in an uncanny way, looked toward a future world where ice would become increasingly scarce as the climate warmed. He noted the decrease in water flowing from shrinking Argentine glaciers—a worry because rivers birthed by those glaciers were the primary source of water for irrigating farmland in the provinces of San Juan and Mendoza.[27] Now, crops in the lowland fields below the once-icy peaks are failing as glaciers vanish and irrigation water becomes scarce.[28]

Today, the loss of alpine ice is clear in photographic comparisons of the same glaciers 100 years ago and today. Geologists have used historical photographs of Montana's Glacier National Park to quantify the rapid disappearance of the park's namesakes and have shown that in the past century, every named glacier in the park retreated.[29] In Alaska, the Mendenhall Glacier is shrinking so quickly that by 2050, tourists enjoying the view from the now-crowded visitor center will see nothing but an iceless fiord. But the loss of alpine ice is about more than scenery. Water pouring from shrinking mountain glaciers accounts for about 20 percent of sea-level rise—almost two inches—in the last fifty years, an amount similar to Greenland's contribution so far.[30]

At the Third Pole, an informal name for the highest elevations of South America, Africa, and Asia where ice and snow cling to mountainsides despite low latitudes, melting is quickly diminishing what little ice remains. Some of the tropical glaciers Thompson cored in the 1980s and 1990s are gone now, including those on the summit of Mount Kilimanjaro and in Papua New Guinea. Today, these glaciers exist only in photographs, as memories, and in the cores packed into Thompson's Ohio State University freezer. For these glaciers, that ice is all we have left for interpreting tropical climates of the past.

Worried about the recent loss of ice and snow fields, one town in Peru sent workers to paint the steep, rocky slopes now exposed by the loss of snow and ice bright white. Their plan was to mimic the reflective and thus cooling effect of the vanished snowpack, hoping to bring back the ice.[31] It was a valiant effort, based on sound physics, but there's no evidence of success because the scale is too small and the warming too great. The Army tried the same approach at Thule in 1953 when it painted the nearly two-mile-long runway white because its dark surface had been absorbing the summer sunlight and melting the permafrost below. Still, some permafrost under the runway continued to melt, leaving behind plane-damaging craters as the thawing soil collapsed.[32]

For ice sheets, there's a deadly feedback loop involving the lapse rate, which is the decrease in temperature with elevation in the lower atmosphere (the troposphere). Measurements and models suggest that Greenland's lapse rate varies between 3.0°F and 5.5°F for every 1,000 feet of elevation gain. That means that as the ice sheet starts to shrink and its surface lowers, the atmosphere around the remaining ice is warmer, further accelerating melting. It's a slow but inevitable death spiral. But the process isn't fast: it takes so much energy to melt a massive ice sheet that even in a rapidly warming climate, many hundred to a few thousand years will pass before the ice covering Greenland completely vanishes.[33]

Beyond the edges of Greenland's ice, satellites clearly show the impact of increasing warmth—expressed as more and greener plants. This phenomenon, called Arctic greening, results from changes in plant type, size, health, and abundance.[34] Its effects are far-reaching. An increase in plants decreases albedo. As well, plants change the movement of water across the landscape and into the atmosphere, and alter the carbon cycle. Some impacts are counterintuitive. As plants grow larger and increase in numbers, the roughness that vegetation creates on the landscape traps more snow from the winter wind. Such snow accumulation can accelerate permafrost melting, as snow cover insulates the ground and prevents it from freezing as deeply over the winter months.[35] The resulting increase in permafrost melting allows ancient soil carbon to degrade and enter the atmosphere as carbon dioxide and methane—another worrisome feedback loop.

The increasing warmth is also threatening thousands of years of human history preserved in the permafrost. As the frozen soil melts, buried hides, bones, and bodies rapidly decay. One scientist estimates that 30–70 percent of the organic remains in Greenland's archaeological sites will degrade significantly in the next eighty years.[36] Carbon-rich Thule, Inuit, and Norse cultural remains will vanish into thin air—oxidizing into climate-warming carbon dioxide as they thaw and decompose. Around the world as Earth's temperature rises, glacial archaeologists are chasing retreating ice. They are finding hundreds of newly exposed artifacts, and even bodies, preserved for millennia, "frozen in time"—for example, Ötzi, the "Iceman," who died crossing the Alps 5,000 years ago, and in Norway, a 3,000-year-old arrow with its feathers perfectly preserved.[37]

SINCE THE EARLY 1990s, Greenland's equilibrium line altitude has been rising as the ablation zone expands. By 2019, the ablation zone had grown 46 percent larger in the north and 25 percent larger in the south.[38] As the climate continues to warm, the ELA will steadily rise toward the U.S. military bases now entombed in the ice. First, it will come to Site 1, which sits at a low elevation near the coast. Then, it will arrive at Camp Century, and finally, at the camp that is highest and farthest inland, Site 2. Once each camp is below the ELA, more snow and ice will melt than accumulates; this net loss will slowly move the ice surface down toward the now-buried camps. It will take years for what remains of the Army's 1950s handiwork to melt out, but eventually, all of it will daylight.

When the B-17 *My Gal Sal* crashed on the ice in southern Greenland in 1942, the abandoned bomber promptly disappeared under the accumulating snow, but as climate warmed and the ice sheet entombing it flowed toward the coast, the firn that covered the plane began melting away. In 1964, the plane reappeared, and in 1995, an expedition retrieved the aircraft, then in pieces, and restored it for museum display. Photographs from the salvage expedition show the center section of the plane perched precariously on a pedestal of ice. By then, the plane was in the ablation zone, and only the shade offered by the plane had protected the ice below from melting.

When similar melting comes to Camp Century, large amounts of a variety of wastes, some quite hazardous, will melt out. The ice around the camp contains millions of gallons of sewage as well as the ethylene glycol coolant, a teratogen that causes birth defects. Toxic PCBs augmented caulking and paints to keep them flexible in the cold. There was lead in the paint and gasoline, and residual diesel fuel and lubricating oil in the generators. Asbestos filled the insulation, and radioactive water poured into the hot waste disposal sump. All told, the military left behind about 10,000 tons of physical waste in the ice sheet at Camp Century alone.[39]

Sites 1 and 2 and Camp Fistclench were each filled with similarly toxic materials, but less of them (about 4,400 tons according to construction records) than at Camp Century, and without the radioactive waste. Hundreds of downed aircraft, broken Weasels, and damaged tractors lie scattered across the ice sheet, left where they failed, many still filled with fuel and oil. Accumulating snow slowly buried much of what the military left behind, which, being mostly in the accumulation zone, became more deeply entombed in ice with every passing year. Scientists and engineers at the time considered the waste to be gone forever. They even hatched a plan to bury nuclear waste in the ice sheet.[40] The radioactivity would have generated so much heat that the waste would have melted to the base of the ice on its own. No one gave much thought to what would happen when the waste got to the glacier's bed—nor did anyone seriously consider climate change and the possibility that the ice sheet might melt away, daylighting all of the waste.

Camp Century was, and still is, in the accumulation zone of the Greenland Ice Sheet. Every year, new snow buries the camp a few more feet below the wind-sculpted surface. As a result, there's no trace of the camp on the surface today. Finding what's left required remote sensing—NASA's Operation IceBridge. For a decade, between 2009 and 2019, specially equipped aircraft crisscrossed Greenland. They used lasers to characterize the topography of the ice surface, and airborne radar to see into the ice and down to the bedrock beneath. That radar found the remains of Camp Century buried under more than a hundred feet of layered ice and snow.[41] The debris and crushed trenches, filled with the remains of the camp, stood out clearly in the radargrams.

Using the IceBridge data as a guide, William Colgan, a geoscientist at GEUS, led an expedition to the site of Camp Century. His team brought along high-resolution, ice-penetrating radar (towed over the ice on sleds by scientists on skis) with which they looked into the snow and ice. They easily identified the crushed camp and found all the trenches shown in the 1960s maps (but not any other, "secret" trenches). The radar identified a Rod well probably containing the still-liquid ethylene glycol antifreeze that condensed steam from the turbine, as well as the sewage sump filled with now-frozen human waste. The camp had not only been buried, but had also been moved—according to their best estimate, a bit more than 14 feet per year to the west-southwest. That's nearly 900 feet closer to the coast in sixty-two years. At that rate, the camp would take more than 35,000 years to reach the ice margin about a hundred miles away—at first glance justifying the Army's 1960s thinking that its debris didn't represent a hazard.

Colgan's team collected numerous snow cores in and around the camp. Their cores consistently found a layer of compact, discolored snow with just a hint of radioactivity—most likely fallout from atmospheric testing of nuclear weapons during the early 1960s.[42] This "occupation layer" resulted from years of compaction by Weasels, heavy swings, tractors, and boots on the snow. It was the icy, buried fossil footprint of the Army's seven-year presence at the site. Tests for iodine-129, a common reactor by-product, showed only low concentrations in all the core samples and no indication of widespread, airborne, radioactive contamination.[43]

Yet an online letter from Franklin to his 1955 classmates at West Point leaves the door open for lingering radiation at Camp Century.[44]

> I can tell you story upon story about opening those snow tunnels and finding "leftovers" from the initial construction of that unique project. And then the residual radiation levels we found were well beyond anything forecast by our stateside experts, so you will have to stretch your imaginations well beyond limits to envision what our crew did to clean the place up. The Chief of the Danish Atomic Energy Commission visited us; a great guy, Lee can tell you more about him. He took his detectors all over the camp looking for residual radiation. His final report

included the words: "I felt the Captain was daring me to find radiation; I did not succeed."

Franklin died of kidney cancer on March 8, 2017—a retired U.S. Army Major General. He was eighty-three. Fifteen years before, surgeons had removed his cancerous prostate and one kidney, but eventually the cancer had metastasized to his lungs and bones.[45] A half century had passed since he decommissioned the Camp Century reactor and made the decision to expose himself, rather than his men, to radiation from the shut-down reactor and the building and trench that encased it.[46]

In 2016, Colgan published a paper with an understated title: "The Abandoned Ice Sheet Base at Camp Century, Greenland, in a Warming Climate." The paper's conclusion attracted a lot of media attention. Colgan's calculations suggested that within the next seventy-five years, as global temperatures rose, Camp Century would no longer be in the accumulation zone. Then, instead of burying the former camp more deeply with each passing year, the snow surface above it would begin dropping—just a bit to start and then, as time went by, faster and faster.

Today, snow rarely melts at Camp Century. Looking forward, as the climate warms, surface melt will happen there with increasing frequency. Eventually, meltwater will percolate into the remains of the camp, and from there it will flow through the ice sheet, carrying whatever pollutants it encounters toward the coast. Emanating from the ice margin, contaminated outwash streams will pour into Baffin Bay, perhaps with an oily sheen of diesel fuel from a crushed generator or leaded gasoline from an abandoned Weasel. One day, a century or two in the future, as melting continues, bits and pieces of Camp Century will daylight on the surface, unroofed as the ice vanishes. If anyone is there to look, perhaps they'll find a once-frozen steak, canned vegetables, a soggy book from the library, some pin-ups, or a leaking barrel of leaded gasoline sticking out of the ice.[47]

Camp Century's future depends, as it always has, on the future of the Greenland Ice Sheet. In the 1960s, scientists and engineers thought that freezing waste in a glacier the size of Greenland meant safe disposal as the ice slowly crept toward the coast. Ice would entomb everything we abandoned, hazardous or not, for what at the time seemed like an eternity. But the world has changed in ways they never fully imagined. Climate change has upended the 1960s view of the ice sheet, just as warm sewage leaking from a pipe beneath Site 2's metal tubes melted the snow beneath the camp and set the station adrift, tilting it to the side and dropping it into the void.[48]

Whenever the remains of Camp Century finally come to the surface, the response is going to be tricky. Someone will need to find tens of millions of dollars to recover the fuel, extract the sewage, pump out the glycol, and deal with any residual radioactivity.[49] It could be that the Danes, who allowed the camp to be built on what was then their territory, will pay the bill, or perhaps the Americans who built the base. The treaty language that enabled American occupation of Greenland and the construction of Camp Century is ambiguous. Maybe both nations will pass their responsibilities off to the Greenlanders, who have re-established independence and on whose land the former base now sits.

Camp Century was once a triumph of man over nature and of American technical prowess during the Cold War. Now, as waste, the former camp is a liability. No one anticipated the environmental significance of a long-abandoned city below the ice or of global warming at the scale and rapidity we now face. The greatest impact of Colgan's paper was not scientific, but social and political. It made climate change real even in one of the world's coldest places.

ICE SHEETS STORE and then release water on a global scale. For several million years, as Earth's climate has swung between glacial and interglacial periods more than fifty times, the same ice sheets have been yo-yoing sea level up and down. The Laurentide Ice Sheet, which in the past covered most of Canada and stretched south and west to New York, New Jersey, Pennsylvania, Ohio, Wisconsin, the Dakotas, and Montana, held about eight times the volume of ice that Greenland's ice sheet holds today. At the last glacial maximum (about 25,000 years ago), land-based ice held so much water that sea level was more than 400 feet below today's average, reshaping the world's coastlines.

Chesapeake Bay, today famous for its crabs and oysters, was then a permafrost-riddled extension of the Susquehanna River valley. There was no English Channel, and people walked freely from Europe to the British Isles across a plain known as Doggerland. Today, trawlers fishing there sometimes bring up freshwater peat and prehistoric stone tools along with their catch. Some believe the first people to inhabit North America came to Alaska from Siberia across a Beringian land bridge that is today below the waves. More than 60,000 years ago, the first Australians probably walked from Papua New Guinea across what is now the Torres Strait. Ice sheets of the past not only shaped many of today's landscapes, but also molded human societies and the genetic inheritance of people around the world.

When the Laurentide Ice Sheet started melting after the last glacial maximum, sea level rose only slowly for thousands of years. Then, the speed of melting increased as the climate warmed. By about 14,600 years ago, the ice sheet was failing fast—the edges were shrinking, the top surface was lowering, and meltwater was pouring into the ocean. Sea level was rising at an unprecedented pace. In just 500 years, it rose almost sixty feet—nearly an inch and a half every year.[50] This speed of sea-level rise could be our future if Greenlandic and Antarctic ice melts at the same rate as ice sheets did at the end of the last glacial period—a reasonable analogy, given how quickly we are warming the planet.[51]

In some parts of the world, land-level fall exacerbates the effects of sea-level rise. The most common culprit is compaction of coastal sediment—the soft, silty muck underlying many nearshore zones. Ten million people live in the island city of Jakarta, the capital of Indonesia. North Jakarta, closest to the coast, is on a river delta and is sinking up to ten inches each year, the result of extensive and unpermitted groundwater pumping and sediment compaction. Forty percent of the city is now below sea level, with concrete seawalls keeping the water out. The situation is so dire that the government plans to move the capital to another island that is well above sea level.[52]

If Jakarta is the most consequential example of rising sea level because of its population, Venice is the best known because of its famed beauty. Italians built Venice on delta sediment similar to what underlies Jakarta, using wooden piles driven into the sand and clay below. These old delta sediments, particularly those rich in organic material, compact over time, slowly settling and sinking. For years, Venetians also pumped groundwater from beneath their city, which made the settling of sediment occur far more quickly than it would have naturally. Today, the pumping has stopped, but the ground has been permanently lowered, and sea level is now rising rapidly as the climate warms. As a result, the city's famous plazas flood with increasing frequency when storm, wind, and tidal conditions conspire to raise water levels for hours to days.

The solution for Venice, albeit a temporary one, is the $8 billion MOSE project, a series of bright yellow gates that close across the openings from the Adriatic Sea to the estuary where the city lies. Each time Venice raises these barriers, it costs hundreds of thousands of dollars.[53] A foot of sea-level rise would mandate that the system be activated every couple of days. As sea-level rise continues, the barriers will separate the lagoon surrounding Venice from the ocean more days than they're connected. The ecological consequences will be devastating, as might be the stench from organic matter decaying in the stagnant water. Fisheries will vanish, along with the livelihoods and lifestyles of those who depend on harvesting seafood. Depending on how quickly sea level rises, the gate system might last fifty years, or at most a century, before it's overwhelmed, along with Venice.[54]

America's Venice might be the coast of Virginia, and especially Naval Station Norfolk, home to the U.S. Navy's Atlantic fleet. The base is big enough to host five aircraft carriers simultaneously, as well as a population of 21,000 people. But its 75 ships, 14 piers, 134 aircraft, and 11 aircraft hangars face an uncertain future. Every five or six years, the water comes up another inch around the docks and the harbor. Slowly but surely, the ocean in southern Virginia is overtaking the land and the Naval Station.

In the century since the Navy established Naval Station Norfolk, sea level has risen about eighteen inches. That means flooded streets, basements full of water, backed-up sewers, and an increasing need to elevate naval infrastructure, including power lines, water pipes, sewers, and roads. The response by people living off base is practical: raise your house above the water on stilts by four feet, seven feet, or even eleven feet.[55] Stilts can cost tens of thousands of dollars or more, but they could make all the difference when the next hurricane, with its accompanying storm surge, arrives. It helps that the state and federal governments often pick up part of the tab for raising a home above the waves.

Making changes to the Navy's biggest shipyard is far more difficult than preserving a private home because politics are involved. Call the work and dollars "climate change adaptation," and a conservative Congress, reluctant to admit that melting ice and warming seas cause sea-level rise, might refuse funding. So, savvy legislative assistants bury funding for rebuilding infrastructure in response to climate change in other projects and verbally disguise such work as a way to "increase resilience." Such double-speak is nothing new. When I was on an external advisory committee to the U.S. National Science Foundation during the Obama presidency, some suggested that we watch our words—"change" was better than "warming," and we never attached "human-induced" to "climate change."

At the time, we did such wordsmithing primarily to avoid the wrath of Lamar Smith, then chair of the House Science and Technology Committee. Smith was from Texas, a state made rich by oil. He spent years trying to remove funding from the Directorate for Geosciences at the National Science Foundation—because this directorate supported a large portfolio of climate research, much of it in Greenland. Some suggested that his approach to climate change and climate science was related to the $600,000 in donations he received over his career from the oil industry, his biggest funder.[56] Smith failed to remove funding from Geosciences and eventually retired. Today, the U.S. National Science Foundation increasingly funds climate-related research, including our work on the Camp Century core.[57] As Lonnie Thompson told *Rolling Stone* in 2005, "Glaciers, have no political agenda. . . . Science is about what is, not what we believe or hope."[58]

Norfolk is not alone—all of the mid-Atlantic states are seeing some of the most rapid rates of sea-level rise anywhere in the world. They can blame the Laurentide Ice Sheet. As we saw with Greenland, as an ice sheet grows, the earth beneath it sinks. In response, outside the edge of the ice, the earth rises, slowly and steadily, over a broad swath of land. This rise is the forebulge. Earth, of course, is no saggy mattress. It's rock. But tens of miles below the surface, beneath the brittle crust and at the top of the more deformable mantle, the rock is softer. It's still rock, but over time this solid material will ever so slowly flow. That slowness of flow is key to understanding why the mid-Atlantic coast is still subsiding today. The forebulge takes tens of thousands of years after the ice is gone to fully dissipate. Today, the bulge is still dropping, and the Mid-Atlantic land surface with it.

In 2015, CNN screamed, "Washington, DC, has a 'Forebulge Problem.'" It was reporting on the research of Ben DeJong, a U.S. Geological Survey employee and University of Vermont graduate student.[59] DeJong split his days between the Vermont clean chemistry labs, where he measured ^{10}Be and ^{26}Al, the same isotopes we used to date the Camp Century sub-ice sediment, and a USGS drilling rig bringing up cores of mud, sand, and cobbles from below

Chesapeake Bay marshland. When DeJong first compiled his data, they were confusing. Scientists know that 60,000–70,000 years ago, global sea level was tens of feet lower than it is today because the Laurentide Ice Sheet was growing across Canada. But cores don't lie. The sediments DeJong examined clearly indicated that saltwater marshes ringed the bay back then at about the same elevation they do today, and he knew that those marshes can survive only if they are within a few feet of sea level.

How could that be? Understanding that Earth is dynamic and changes shape slowly over time is the key. Sea level was low 60,000 years ago, but so was the land around the bay. The forebulge from the prior glaciation ending about 130,000 years ago had collapsed. Comparing a sediment deposit's elevation today with global sea level in the past wasn't going to work. DeJong needed to think about how the land around the bay moved up and down over time. So, too, do urban planners, because today Washington, DC, is subsiding rapidly. America's capital city is feeling the effects of the decaying forebulge from the last advance of the Laurentide Ice Sheet. DeJong's data suggest that its collapse will continue for a few tens of thousands of years into the future.

Forebulge subsidence, combined with sea-level rise and sediment compaction, is hitting the mid-Atlantic states with a triple whammy. It's the reason that these states, including Washington, DC, and coastal Virginia, are seeing local sea-level rise at rates faster than much of the world. The outcome isn't good. Today, water pushed up the Potomac River estuary by rising sea level is slowly inundating the Tidal Basin and its iconic cherry blossoms.[60] At high tide, floodwaters often lap onto the walkway around the basin. Some cherry trees, their roots saturated, have already died. As the sea continues to rise, water will eventually come to the capital's other attractions, dampening Jefferson, King, and one day, Lincoln and leaving the Washington Monument an island.

Once sea-level rise in and around Washington exceeds a few feet, the problems, and the costs, will multiply. Metro stations and tunnels will flood, major roadways will become impassable, and floodwaters will inundate many of the museums on the Mall. All of this will probably happen in the next century and cost tens of billions of dollars to address.[61]

FOR THOSE NOT living on the shore, the effects of climate change don't directly involve sea-level rise, but do, in part, originate in the Arctic. Consider Texas. In June 2023, a heat dome—driven by a massive area of high atmospheric pressure—settled over the state. A large meander in the jet stream blocked the high-pressure system and its associated heat from moving eastward. The cloudless, stagnant air allowed the ground beneath to warm under the baking low-latitude summer sun. Day after day, the atmosphere over Texas grew warmer.

Temperatures on the ground steadily rose until they were well over 100°F. Heat indices (the amalgamation of heat and humidity) made it feel 10°F–20°F warmer. El Paso had thirteen days with the mercury over 100°F. The city of Midland stayed that warm for fourteen days. All-time high-temperature records fell over much of the state, where the daily weather history extends back well over a century. A month later, a similar heat dome covered southern Europe. Days of searing temperatures in Italy, Greece, and Spain peaked at over 113°F, breaking the all-time heat record in Spanish Catalunya. Then, on July 16, 2023, China broke its all-time high temperature record by 3°F when the mercury hit 126°F in Sanbao, a small agricultural town in northeastern China.

Atmospheric scientists have argued that climate change appears to make such "omega blocks" (so named because the shape of the meander resembles that Greek letter) more common. Many suspect that Arctic warming—specifically, the change in albedo driven by the loss of sea ice, a diminished snowpack, and the shrinking Greenland Ice Sheet—is the driver. There's good evidence to believe that the lessening difference in temperature between high and low latitudes encourages the jet stream to meander, which facilitates the formation of heat domes.

The Texas heat dome is one of many that North America has experienced recently, and it won't be the last. In 2021, a similar high-pressure-driven heat dome settled over usually temperate Washington, Oregon, and British Columbia. Thermometers soared to 116°F in Portland. All over the Pacific Northwest, cities broke long-standing daily records by 5°F, even 10°F. Heat-related deaths shot up, roads buckled, and people with symptoms of heat exhaustion packed emergency rooms. Statistically, the heat dome was a one-in-ten-thousand-year event. Various approaches to attribution—the science that estimates the likelihood that human-induced climate change makes extreme events more likely—concluded that, yes, our changing climate was in part to blame not only for the extensive heat dome, but for the extreme high temperatures.[62]

People, however, don't feel the effects of extreme heat uniformly. Social status matters. In the United States, wealthy neighborhoods have more trees, more shade, and cooler temperatures than poorer neighborhoods.[63] And consider that many prisons in Texas, one of the hotter states in the union, are not air conditioned. At least nine incarcerated people died from the heat during the June 2023 event, despite lawsuits claiming that such conditions are cruel and unusual punishment. Over the past twenty years, researchers suggest that heat caused 13 percent of prison deaths in Texas as interior temperatures in some facilities peaked at over 140°F.[64]

In every heat wave, the unhoused population suffers inordinately—because they often have little access to cooling, and many have untreated, chronic medical conditions. In contrast, by 2020, nearly 90 percent of American homes reported having some sort of air conditioning, as compared with 2001, when just over 75 percent of homes had cooling systems.[65] In those nineteen years, global average atmospheric temperature rose about 1°F, and average annual temperature rose almost everywhere on the planet—

with the exception of that patch of the North Atlantic just south of Greenland and the Southern Ocean off the tips of Africa and South America—places where no one lives.

Although the evidence is weaker, some argue that our warming climate also drives the mirror image of heat domes: winter's arctic blasts. They, too, happen when the jet stream meanders and loops, plunging toward the equator, allowing frigid air to pour out of the polar regions and into lower latitudes. The most recent and best-known example is the Great Texas Freeze of February 2021, which had the entire state under a winter storm warning, left 4.5 million households without power, and killed more than 200 people.[66] Although it seems counterintuitive, as the climate warms, short-lived bouts of extreme cold are likely to continue along with the loopy jet stream.

And then there are the increasing storms. Catastrophic hurricanes are now becoming the norm. Between 1979 and 2017, the number of strong hurricanes (categories 3, 4, and 5) increased substantially.[67] The immediate cause was sea-surface temperature, which rose—like the atmospheric temperature—almost 1°F over the same time period.[68] Hurricanes are heat engines. They derive their power from moist tropical air and the warm seawater below. That's why these storms are a summer phenomenon, and it's why a warmer ocean drives stronger storms. The future looks no better. Atmospheric models consistently show that the frequency of Category 5 hurricanes, the most damaging storms, is likely to increase as Earth continues to warm. Those storms will last longer, travel farther north, and rain harder.[69]

Warm water at the ocean surface also causes rapid intensification of hurricanes. In August 2023, it took Hurricane Idalia less than twenty-four hours to go from a weak Category 1 hurricane near the western tip of Cuba to a devastating Category 4 storm just before it slammed into western Florida. In that short time, the storm's average wind speed increased by fifty-five miles per hour—an increase fueled by exceptionally warm ocean water. By July 2023, surface-water temperatures in the eastern Gulf of Mexico were nearly 2°F above average and breaking records. Hurricane Harvey should have been a warning when it devastated southern Texas in 2017. Waters in the eastern Gulf of Mexico were exceptionally warm that summer, too, increasing evaporation rates and feeding moisture into the storm, which intensified in less than twenty-four hours from a tropical storm to a Category 4 hurricane. When Harvey stalled for several days, it dumped five feet of rain on parts of Houston.[70] That water came right out of the nearby Gulf, readily evaporating from its overheated surface.

As we lose the snow and ice of the Arctic, the world's oceans will continue to warm along with the atmosphere. As a result, storms will get stronger, and rainfall heavier, because a warmer ocean increases evaporation, and a warmer atmosphere holds more water vapor than a cooler atmosphere. Rapid intensification of storms will become more common, shortening the window for evacuation. Stronger storms generate higher storm surges, and that water, coming atop more than a century of sea-level rise, will cause currently unthinkable amounts of damage.

In 1992, Hurricane Andrew came ashore just south of Miami as one of only four Category 5 hurricanes known to have struck the continental United States. The accompanying storm surge pushed the ocean surface seventeen feet above normal levels as peak winds reached 175 miles per hour.[71] The combination caused $50 billion worth of damage. By 2018, sea level was about three inches higher. That number sounds trivial, but Swiss Re, a firm that insures insurance companies and thus must take climate change into account to set rates and remain solvent, suggests that even that few inches of sea-level rise could nearly double average annual losses from hurricane-driven flooding in the Miami area.[72]

Henri and Adele Bader's first house in Miami was a thousand feet from the beach and ten feet above sea level.[73] Ten feet of sea-level rise requires melting of less than half the volume of the Greenland Ice Sheet, well within the realm of possibility when the ice below Camp Century is gone. NOAA suggests that if carbon dioxide concentrations in the atmosphere continue their unabated increase, sea level could rise seven feet by 2100. Such rising waters, along with warm temperatures, might destabilize ice sheets by floating outlet glaciers off their beds, reducing friction, and thus quickening ice flow toward the ocean. If that happens, the resulting ice sheet collapse would double the expected sea-level rise and put most of Miami, including Bader's former home, underwater by 2100.[74]

Later in life, the Baders moved closer to the water but farther above sea level. Their eleventh-floor apartment had a view over Biscayne Bay.[75] That was a good move in terms of avoiding rising seas, but a bad move when Hurricane Andrew arrived. The storm left the elderly couple stranded in their apartment for days. It blew out their windows, cut their power, and left their apartment's water-soaked carpet covered in broken glass. They survived the storm by hiding in a side room. Both were too frail to walk down eleven floors of stairs for help, and they lived on what food and drink they had in their apartment when the storm hit. Rain destroyed Henri's computer, on which, even in retirement, he continued to do the mathematics that he loved.

Henri died six years later, in 1998 at age ninety-one, his lungs compromised by emphysema from decades of smoking.[76] Adele passed in 2001. Friends scattered their ashes together at sea, a fitting place for a man whose life focused on frozen water, some of which by then was flowing into the global ocean and lapping ever higher on Miami's beaches.[77] It seems that Wally Broecker, a geochemist and for decades one of the most influential climate scientists, had it right when he said, "Climate is an angry beast, and we are poking at it with sticks."[78]

DAVID FARRIER, a professor of literature at the University of Edinburgh, reflected on the cultural events recorded in the world's ice—plagues, smelting, the advent of fossil fuels, the Green Revolution, and the COVID pandemic. To him, ice sheets are a library of our shared climate history.

> Spending time in the library of ice reminds us that our history is bound up with that of the planet. As that library comes under ever increasing risk, we should remember the fate of another great library. Legend tells that the Library of Alexandria burned to the ground, but the truth is less spectacular. As the Roman Empire fell into decline, people simply neglected to protect and preserve the fragile papyrus manuscripts that were stored in the Library of Alexandria. Gone with it were the greatest treasures of the ancient world: hundreds of years of civilizations' stories, memories, knowledge, and wisdom. The greatest library in history was lost to neglect. Unless we act now, the library of ice will meet the same fate.[79]

Ice is by nature fragile, and as Earth warms, our library of climate history is flowing away in murky streams of glacial meltwater and into the world's oceans. Soon, racks of cores checked out from Farrier's ice-sheet libraries may be all we have left, analogous to the surviving fragments of Greek tragedies from the Library of Alexandria, incomplete but tantalizing reflections of their playwrights' greatness.

Breaking the ice waste of the north—its ice sheets, sea ice, permafrost—and releasing its secrets has had consequences beyond what most people in the 1950s could ever imagine. The foundation of cold-region science established by the U.S. Army on the frozen landscapes of Thule, Tuto, and Camp Century—snow studies, ice tunnels, and deep ice cores—fundamentally altered how we understood polar regions. Over and over, science showed that the Arctic was particularly fragile and vulnerable to change. During the decades that followed, building on these data, we learned just how important the far north and its ice were to maintaining Earth's global climate equilibrium.[80]

Now, handfuls of sand and mud from the bottom of Camp Century's ice core leave us an equally poignant message—although it took almost sixty years to decipher. The Greenland Ice Sheet is fragile. Much of it vanished in the past when temperatures were similar to today's and carbon dioxide concentrations in the atmosphere were far lower. We now know that unless we change our ways, Greenland's ice is likely to melt again, and when it does, the consequences for people and society will be immense.

In 1960, it was easy to ignore the climate and climate change. Keeling's curve was two years old. Atmospheric carbon dioxide was only at 315 ppm—a mere 35 ppm above the typical interglacial concentration. Now, Keeling's curve has risen steadily for sixty-four years, and as of 2024, there are more than 420 ppm of carbon dioxide in the air we breathe. Ignoring climate change dooms Greenland's ice and constrains us to a future in which Greenland will become green, and what was once snow and ice will become ocean water. Unless we act, every coastal city around the world will be underwater, and the outlines of continents will change, leaving people looking for higher and higher ground.

In 2012, fiction writer Emma Donoghue acknowledged that in writing historical fiction, "you will also be haunted by a looming absence: the shadowy mass of all that's been lost, that can never be recovered."[81] Donoghue's words apply just as well to our knowledge of Greenland and its ice. Using multimillion-dollar atom counters as mass spectrometers, we can tease out bits and pieces of the past. We can listen to isotopes telling cryptic stories, and we can interrogate frozen tundra twigs, but the stories that result lose so much over time and in translation. Even with all these new data, the best we can do is to sketch the broad outlines of what Greenland looked like as Earth warmed in the past. These outlines, fuzzy as they are, are sobering. We know now that much of the Greenland Ice Sheet, which defines the Arctic, melted away at least once before.

Donoghue's "looming absence" also overshadows Earth as we knew it. Within a human lifetime, our coastal cities will begin an unstoppable transition to no more than shadowy, submerged masses. Entire cultures, experiences, and ways of life will vanish.

For many, winter's long nights will come without the usual cold as climate zones march northward. Snowballs and ice skating will be a memory for many, even at high latitudes. Low-latitude cities will slowly become uninhabitable as temperatures increase, and farm fields will lie fallow as droughts desiccate soils and rivers run dry. Wildfires will burn through forests, grasslands, and subdivisions as monster hurricanes ravage coastlines inundated by rising seas. It's a dystopian future hinted at by the summer of 2023.

In a 2016 paper, Peter Clark, a geologist who studies the dynamics of glaciers and ice sheets, and the research team he led considered the consequences of modern-day policy for climate change and Earth over the next 10,000 years.[82] Their conclusions were both eye-opening and frightening: even if carbon emissions ended tomorrow, concentrations of carbon dioxide in the atmosphere would decline only slowly, leading to inescapable changes in Earth's climate, including warmer global temperatures, melting ice sheets, and rising sea levels.[83]

Left to its own devices, the carbon dioxide we have emitted will slowly leave the atmosphere. Plants will take it up, as will soils. Much will dissolve in seawater. Minerals, as they slowly weather, will contribute elements that will combine with the carbon and thus isolate it from the atmosphere. This disappearing act will take many thousands to tens of thousands of years. Most people don't appreciate the stubborn persistence of carbon dioxide and the slow rate at which natural processes transfer it from the atmosphere to the oceans, the biosphere, and the solid Earth. It's the reason why what we do today matters for the next thousand generations.[84]

Clark suggests that no matter how quickly we reduce emissions, global average temperatures will remain higher than they have been during the Holocene for at least the next 10,000 years—unless we can remove carbon dioxide from the atmosphere. Even in a scenario where carbon emissions fall rapidly, sea level will rise inexorably for thousands of years, topping out up to 90 feet above today's shoreline as much of Greenland's and Antarctica's ice melts. If we continue emitting carbon dioxide at current or increasing levels, the impact will be even more profound. Sea-level rise could exceed 170 feet, with the rate of rise exceeding 10 feet in a single human lifetime—a stunning rate of change, going beyond even the most rapid rise at the end of the last glaciation. Over millennial time scales, climate change will literally reshape the geography of earth and sea so that most coastlines will become unrecognizable.

Amid today's climate crisis, it's difficult to understand a culture that put a military base inside an ice sheet and then warmed it with hundreds of thousands of gallons of diesel fuel. Perhaps people in 2100 will look back and think much the same of us. How could we have ignored the growing signs of a changing climate such as rapidly melting ice sheets, or let cities like Venice, with their thousand-year history and grand architecture, go beneath the waves? Yet, when we look back to Greenland, there's a glimmer of hope. The Army did have the foresight to design and implement nuclear power, realizing that using fossil fuels in remote areas wasn't sustainable. Of course, more than a half century later, the accumulation of waste from the world's atomic power plants remains a largely unsolved problem.

IN 1972, BADER nominated Wasserburg for the Day Medal, one of geoscience's highest honors. Bader's nomination was short, funny, and self-deprecating. Wasserburg's response was humble, but more than fifty years later, it carries a message that gives me hope for our planet even as the climate dramatically warms.

> The future of our profession, our real hope for advancement, will probably not be found in this hall here tonight filled with many dignitaries, but rather around the corner in an inexpensive bar or restaurant where there will be a half-dozen or so very bright young people drunk on the excitement of science and full of energy and imagination.[85]

It's exactly this forward-looking, confident attitude that brought us Camp Century and its ice core—a core that in 1969 told Dansgaard the secrets of a thousand centuries of climate history and now warns us about the fragility of Greenland's ice sheet. My generation has so far lacked the will and imagination to do what it takes to avoid the worst effects of a warming world. The next generation has little choice but to do better. Today, we need every one of those very bright young people and their creative energy to help 8 billion citizens of Earth cope with, and hopefully reverse, climate change.

Thule, Tuto, Fistclench, and Camp Century remain so far out of the ordinary that more than a half century later, they still captivate us and our imaginations. It's easy to think of Camp Century and its ice core in isolation. But that misses the point. Environmentally impactful, resource intensive, sexist, and colonial, these camps and the science they catalyzed remain examples, however flawed, of what people can do when they are determined to solve a problem. In the 1950s and 1960s, scientists and engineers figured out not only how to survive in the world's most extreme climate, but how to thrive, while at the same time developing an understanding of the life-threatening environment in which they worked.

Climate change is a "wicked problem par excellence," a series of linked problems impossible to solve in isolation, not unlike military operations in the Arctic during the 1940s and 1950s.[86] That gives me hope that if we can apply the same determined attitude and muster the intellectual and financial resources that Bader, Bender, Hansen, Ueda, and the U.S. military did then, we have a chance to address climate change and its impacts now. Of course, we've barely started down this path, and climate change is a far greater challenge because fossil fuels underpin our current global economic system.

Although time is running short, we can still preserve Earth's most frigid places—those that hold in their deep blue ice the memories of Earth's climate over a time uniquely meaningful to us as

The Knoxville News-Sentinel Sunday, Aug. 23, 1959 Page B-3

A cartoonist's view of Camp Century soon after construction began. On August 23, 1959, this cartoon accompanied a story in the *Knoxville News Sentinel* titled "Offers Year-Round Comfort—U.S. Building A-Powered Town under Greenland Ice Cap." © *Bill Dyer—USA TODAY NETWORK.*

a species.[87] Camp Century's permacrete recalls when the earliest humans roamed the savannahs of Africa. The ice knows when people first left that southern continent and migrated to a moister and greener Middle East as the climate cooled and ice sheets expanded around 100,000 years ago. Toward the top of the core, the ice remembers when we settled in cities and first smelted lead and silver. And the ice recalls when we began to fundamentally alter Earth's atmosphere.

To preserve Greenland's ice sheet, and keep the ice under Camp Century, we must act today to stabilize Earth's climate. The critical first step is rapidly decarbonizing the global economy. We need to do that now in order to limit the total amount of carbon added to the atmosphere—the primary determinant of how warm Earth will get and for how long. But decarbonization is not enough to save

Greenland's ice. Unless we lower the concentration of atmospheric carbon dioxide and thus cool the planet, the melt will go on.

Right now, the technology to pull carbon out of the atmosphere at a meaningful scale doesn't exist—but neither did effective ice-coring technology in the 1940s. We can make a dent in atmospheric carbon concentrations with better soil management techniques and by planting trees and protecting forests. But such ecosystem-based carbon removal approaches are subject to the vagaries of the changing climate itself. There's little certainty about how temperature and precipitation over centuries will affect long-term carbon storage.[88]

Almost sixty years after Hansen and his team finished drilling, Camp Century's ice core continues to tell stories that each day grow more relevant as Earth moves quickly toward a climate state that modern humans have never experienced. Nature has already run the experiment in which the planet stays as warm as it is today for many thousands of years. Camp Century's permacrete and the frozen ecosystem under the ice told us the result—a shrunken ice sheet and sea levels high enough to swallow many of our coastal cities. The lives of so many people and the existence of so many places on our planet depend on our listening and learning from the ice and the frozen sediment beneath, and then hearing what they have to say: "Change your ways."

We should consider ourselves warned. Greenland's ice vanished before. It's time to act collectively and decisively and stop burning fossil carbon before the ice is gone again.

ACKNOWLEDGMENTS

There are so many people to thank in so many different ways. My family first and foremost, Christine, Marika, and Quincy, and my mother Babette, for putting up with my years-long absence as I read, wrote, and babbled on at dinner about the most interesting new tidbit about the Cold War I'd discovered. Because it was COVID time, I didn't go far, but spent many days hidden away in my third-floor office and in the corner of our breakfast nook.

The folks at W. W. Norton made this book happen. Amy Cherry found me, shaped my proposal, and took the textbook writer out of me with good humor, Track Changes, and an amazing way with words. Together, she and Huneeya Siddiqui did the editing work that took my three rambling drafts full of deep dives into icy trivia and shaped them into a book. Thanks to you both, and to Norma Roche, whose copyedits clarified my writing and made this book far more consistent and readable.

Colleagues at the University of Vermont have been instrumental in this project. Initial and ongoing support from Taylor Ricketts and the University of Vermont's Gund Institute for Environment were catalytic not only for this book, but for the Oldest Ice Workshop in 2019 that kick-started this entire project. Particular thanks to Christine Massey for organizing that three-day science blitz. Nico Perdrial, a University of Vermont geochemist, was an important sounding board not only for science but for history, as he translated articles from French to English when my poor understanding of other languages got in the way.

Over thirty years, graduate and undergraduate students at UVM pushed me to think differently and to question assumptions. They include more than forty graduate students whom I have advised and many thousands of undergraduates in my classes—all of whom told me, loud and clear, the importance of making science interesting and relevant.

Special thanks to Cat Collins, Juliana Souza, and Halley Mastro. As graduate students at UVM, these three have worked tirelessly to understand the Camp Century sub-ice sediment, each bringing their own special skill set to a team effort. Their enthusiasm, inquisitiveness, and excitement for things old, cold, and Gouda is infectious. Mary Kueser joined us in her senior year at UVM and produced a podcast about Camp Century, *Frozen History*.

Josh Brown, the University of Vermont's environmental writer, has had an outsized impact on me and my writing. For years, he's joined my graduate seminar in writing and helped me and the students become better communicators. I had the honor and pleasure of spending two weeks in Greenland with Josh. Basil Waugh from the Gund Institute helped me hone my communication skills further with his mantra: remember your talking points, stay focused, and don't get distracted. Despite that great advice, I can always find another rabbit hole to dive down.

Support from the faculty, students, and especially the leadership team of the University of Vermont's Rubenstein School of the Environment and Natural Resources has been so valuable, most especially the encouragement of Deans Mathews, Erickson, Pontius, and Strong, who saw value in the public communication of environmental science to the broader community. Tom Visser encouraged my lifelong interest in history (and now book writing) with an enthusiasm for all things old and interesting. John Hughes, a friend and mentor since I met him thirty-one years ago, is the reason we have our cosmogenic nuclide lab and can do the science that helped illuminate Camp Century's history. Lee Corbett not only managed the lab to produce the absolute best data, but has been a constant sounding board and enthusiastic reader of all the Thule trivia I find.

Many European scientists have been critical to the research that underlies this book. My deepest thanks to Dorthe Dahl-Jensen and Jørgen Peder Steffensen, without whose careful storage of the Camp Century core we would never have been able to do this science. Conversations of many years with both of them have opened my eyes to the wonders of ice-core science. Paul Knutz showed me the stories told by old till, a critical stepping-stone to our work together on the Camp Century core. Ole Bennicke generously shared his time and knowledge of ancient Greenlandic plants. Pierre Henri Blard and Jean-Louis Tison spent days in the freezer logging and cutting the Camp Century sub-ice core, much of it training, and then assisting, Drew Christ in the month-long job of cutting the frozen material. Drew and I shared much wonder as this project evolved. His meticulous work cutting the core and doing the initial core processing is why so many people have so much data today.

To the people who, over my forty years as a scientist, mentored me along the way, not only in doing science but in communicating science, I owe you all a huge thanks. Thom Davis introduced me to lake coring and fossil finding, which has come in so handy. Thanks Thom. David Dethier instilled in me a love of all things glacial while insisting that I write clearly and always in the active voice. He and other members of the Williams College Geology Department have supported me since I took my first geology class, Lakes and Oceans, in the fall of 1980. Rolfe Stanley, another Williams graduate and a mentor to me at Vermont, taught me to double-check everything I did and never work in a vacuum—always ask for and respect feedback from others. His advice has guided my work since I came to Vermont in 1993. I wish he were here to read this book.

At the University of Washington, Alan Gillespie introduced me to paleoclimate, the Sierra Nevada, and western glacial geology while suggesting that I never split an infinitive. His advisor was

Gerry Wasserburg. It's a small world, but Gillespie had no idea Wasserburg had ever set foot on a glacier, much less helped drill and analyze the first American ice core. Marc Caffee, John Southon, Bob Finkel, Dylan Rood, and I spent countless nights awake watching accelerators spit back ^{10}Be and ^{26}Al data. They taught me what it means to perform analyses that are accurate and precise, and to push the limit of what people can do with a machine and a lab.

I treasure the weeks of fieldwork that I did in Greenland with Jeremy Shakun, Eric Portenga, Lee Corbett, Josh Brown, and Joseph Graly. None of that would have happened without the support and encouragement of Tom Neumann, who showed me the ropes of working on the island during our first trip together in 2008. He inadvertently taught us all to read the label carefully when buying what by the appearance of its label should have been beer when in a remote Greenlandic town on a Sunday afternoon.

Dave Montgomery, with whom I've written for almost two decades, gets a special thanks for telling me, of course you can write a trade book, you've written two textbooks. Michael Mann, Richard Alley, and Bill McKibben have been sounding boards for science and action over the years and inspirations for how to communicate climate science outside of academia. Special thanks to all of you for your advice on teaching, research, and negotiating the wild world of bringing climate science to the public eye. Cyd Slayton shared her knowledge and passion about early climate change scientists and Ron Doel provided invaluable assistance and encouragement regarding arctic environmental history.

I had so much help finding obscure material from so many people. Trina Magi, a superb UVM research librarian, introduced me to critical resources that I had no idea even existed. The UVM interlibrary loan staff found obscure reports in the oddest places. Brendan McDermott, thesis/dissertation coordinator, Joel Sparks, laboratory manager, and Lawford Anderson, professor of earth science, all helped me navigate Boston University history and figure out Langway's early years. Chris Burns and Prudence Doherty at UVM's Special Collections helped us set up our online image archive (Iceworm—Camp Century and the U.S. Military/Science Complex, https://campcentury.omeka.net/), which has been critical to learning and communicating about the Arctic decades ago. Simon Beck taught me about the Flying Boxcars that were so critical for resupplying early expeditions as well as Sites 1 and 2. Frederick Stoss and Jessica Hollister at the University at Buffalo library helped with Langway archives, while Laura Kissel, polar curator at the Ohio State University, found photographs in its archives I'd never seen before. I thank Georgina Ward at the University of Melbourne and Christian Salewski at the Alfred Wegner Institute for finding and providing historic images.

And a huge shout-out to all those people I will never know who in different ways made the research underlying this book possible. Thanks to everyone in the Department of Defense who scanned once-secret and now-declassified documents and made them publicly available with a few keystrokes when COVID locked down libraries and archives. Thanks to the bloggers and webmasters who filled the web with stories, photographs, and tidbits of information that no library holds. To folks who sold me musty ephemera on eBay, thank you for listing items I'd not otherwise been able to find. You provided a 1950s and 1960s window into Greenland, the Cold War, Camp Century, and its ice core like no other.

Although many people involved with Camp Century have now died, I spoke to a few who were there, including Jon Fresch, the camp photographer, who scanned and sent me a spectacular set of photographs; and Bill Vest, who worked for several years with Langway processing cores under the ice. I learned so much from speaking to people whose fathers had been in Greenland and at Camp Century. Dagmar Cole told me many stories about her father's experiences, as did Susan Henderson King, who sent me color photographs of her father on the ice. I also had long conversations by phone, email, and in person with Lucy Bledsoe and Nancy Kates about the 1950s, gender, and science. I value their friendship and advice as perhaps the only two other people as fascinated with this place and period of time as I am.

But the oddest connection was with Hans Wittich, with whom I'd gone to high school. I don't think we'd seen each other since graduation in 1980, but in just a few hours he answered an email from me asking if by any chance he was related to Joe Franklin, who pulled the reactor from Camp Century. He was. Franklin was Hans's uncle, his mother's twin brother. Thanks, Hans, for making that connection and telling me more about Franklin.

Lonnie Thomson took hours on the phone explaining his interactions with Langway and the science he did on high-altitude ice cores. My deepest thanks go to John Fountain, who explained in detail the University at Buffalo Geology Department in the 1970s–1980s. Julie Palais and H. Jay Zwally were exceptionally patient and helpful, spending time by email and on the phone discussing the 1970s, 1990s, and 2000s at the U.S. National Science Foundation.

This book, and the science of climate and landscapes that I've been able to do, would never have happened without three decades of continuous support from the U.S. National Science Foundation. Since 1994, the NSF has supported my students, our laboratory, and me to do all sorts of science, much of it related to the Arctic, to Greenland, and to Earth's climate and the resulting landscape response over time. I give special thanks to Marc Stieglitz and Henrietta Edmonds, who supported and guided our work in Greenland since 2008, as well as Russel Kelz and David Lambert, who oversaw grants supporting our cosmogenic nuclide laboratory at the University of Vermont.

This book, and my experiences shared in it, would not have been possible without a series of nearly thirty National Science Foundation grants, all of which supported either fieldwork or laboratory analyses. The discoveries they catalyzed made much of the story I have told here possible. In particular, the work reported in this book resulted from support by the following grants: NSF-EAR

9702643, 0713956, 1023191, 1103381, 1602280, 1733887, 1735676, 2116209, and 2300560. Of course, the ideas expressed in this book are mine, as are the errors that I have inevitably made trying to look back in time.

Public funding to support scientific and climate research in Greenland is integral to the story I've done my best to tell. Without decades of U.S. military support, there would have been no Camp Century and no under-the-ice coring. Without the U.S. National Science Foundation supporting ice-drill development in the 1960s and then the ice-core storage facility, first at Buffalo and then in Colorado, there would be no Camp Century sub-ice core to analyze. Denmark's support of its ice-core storage facility and ice-core researchers kept the sub-ice sediment frozen for the last thirty years. To everyone who paid taxes and supported these endeavors, thank you for saving those frozen bits of soil that speak to us about the past and thus inform our collective future.

NOTES

PROLOGUE

1. Eve Savory, "Wallace Broecker: How to Calm an Angry Beast," CBC News, November 19, 2008.

CHAPTER 1: THE INHOSPITABLE ISLAND

1. M. Thomas et al., "Paleo-Eskimo mtDNA Genome Reveals Matrilineal Discontinuity in Greenland," *Science* 320, no. 5884 (June 27, 2008): 1787–89, https://doi.org/10.1126/science.1159750.
2. Canadian Museum of History, "Disappearance of Dorset Culture," in *Lost Visions, Forgotten Dreams, Life and Art of an Ancient Arctic People*, exhibition, Canadian Museum of Civilization, October 17, 1996–May 19, 1997.
3. The National Museum of Denmark, "Prehistory of Greenland," https://natmus.dk/organisation/forskning-samling-og-bevaring/nyere-tid-og-verdens-kulturer/etnografisk-samling/arktisk-forskning/prehistory-of-greenland/.
4. The terrestrial age, or landing age, of a meteorite is determined by measuring its concentration of different isotopes (including ^{10}Be and ^{26}Al), all formed by cosmic radiation while the meteorite spent billions of years in space. It's simplest and most reliable to calculate an age for a small meteorite, in which there is little change in cosmic-ray intensity with depth. When large bodies like Saviksue break up, most pieces are not from the surface and thus have extremely low concentrations of these cosmogenic isotopes. This makes terrestrial ages difficult to calculate reliably. Cape York data are reported in Thomas Smith et al., "The Constancy of Galactic Cosmic Rays as Recorded by Cosmogenic Nuclides in Iron Meteorites," *Meteoritics & Planetary Science* 54, no. 12 (December 2019): 2951–76, https://doi.org/10.1111/maps.13417. The authors assign a terrestrial age of 67,200 ± 406,000 years, which is so uncertain as to be meaningless.
5. DNA data show that ivory sourced in Greenland is found throughout Europe, indicating the breadth and distance of the Norse trade routes. The data are provided by Bastiaan Star et al., "Ancient DNA Reveals the

Chronology of Walrus Ivory Trade from Norse Greenland," *Proceedings of the Royal Society B: Biological Sciences* 285, no. 1884 (August 15, 2018): 20180978, https://doi.org/10.1098/rspb.2018.0978.

6. The DNA evidence suggests mixing between European men and Inuit women, but not between Inuit men and European women. See Ida Moltke et al., "Uncovering the Genetic History of the Present-Day Greenlandic Population," *American Journal of Human Genetics* 96, no. 1 (January 2015): 54–69, https://doi.org/10.1016/j.ajhg.2014.11.012.
7. There is a long debate over why the Norse settlements in Greenland eventually failed. A recent, informative, and readable review is provided by Eli Kintisch, "The Lost Norse," *Science* 354, no. 6313 (November 11, 2016): 696–701, https://doi.org/10.1126/science.354.6313.696.
8. Marisa Borreggine, Konstantin Latychev, Sophie Coulson, Evelyn M. Powell, Jerry X. Mitrovica, Glenn A. Milne, and Richard B. Alley, "Sea-Level Rise in Southwest Greenland as a Contributor to Viking Abandonment," *Proceedings of the National Academy of Sciences* 120, no. 17 (April 25, 2023): e2209615120, https://doi.org/10.1073/.
9. Moltke et al., "Uncovering the Genetic History of the Present-Day Greenlandic Population."
10. The story of the German expedition is well told by Jon Gertner, *Ice at the End of the World: An Epic Journey into Greenland's Buried Past and Our Perilous Future* (New York: Random House, 2019).
11. "Archive of a Refugee Scientist," Archives and Special Collections, University of Melbourne, June 9, 2016, https://blogs.unimelb.edu.au/librarycollections/2016/06/09/archive-of-a-refugee-scientist/.
12. "Greenland Inland Ice Weather Stations," in *Encyclopedia Arctica*, vol. 7, *Meteorology and Oceanography* (Digital Collections, Dartmouth College Library), reports the following temperatures in the Eismitte snow cave: on the floor, −4°F; on the level of the table, 14°F; near the ceiling, about 21°F.
13. "Greenland Inland Ice Weather Stations."
14. Sorge presents the data from his snow pit, including his interpretations—the first data set of its kind—in Ernst Sorge, "The Scientific Results of the Wegener Expeditions to Greenland," *Geographical Journal* 81, no. 4 (April 1933): 333, https://doi.org/10.2307/1785439.
15. Henri Bader extended the work of Sorge and added a more mathematical touch in Henri Bader, "Sorge's Law of Densification of Snow on High Polar Glaciers," *Journal of Glaciology* 2, no. 15 (1954): 319–23, https://doi.org/10.3189/S0022143000025144.
16. The stories of this couple, and several other people who died high on glaciers many years ago and are now melting out, are presented, along with photographs, by Philip Oltermann and Kate Connolly, "Melting Glaciers in Swiss Alps Could Reveal Hundreds of Mummified Corpses," *Guardian*, August 4, 2017, World News. With climate warming, expect more of these bodies to appear as glaciers around the world shrink.
17. An entertaining description of the Snowball Route and the crashes and rescues, especially the series in East Greenland in 1942, is provided in

Stephan Wilkinson, "Seven Down: Courage during an Air Rescue Disaster," *Navy Times*, December 3, 2019.

18. Paul W. Tibbets, Clair C. Stebbins, and Harry Franken, *The Tibbets Story* (New York: Stein and Day, 1978).
19. Details, including coordinates of the sites at which the planes crashed and photographs of two of them, are provided in "Greenland Air Crash (1942–1945)," Warcovers, https://www.warcovers.dk/greenland/crash_list.htm. A more narrative description is provided in Jack Dorsey, "B-17 Frozen in Ice and Time for 53 Years May Fly Again," *Virginian-Pilot*, October 1, 1995.
20. The other two had quite different endings. The *Alabama Exterminator* crash-landed on the rocky western Greenland coast. The crew walked away, and native Greenlanders salvaged the metal fuselage for "bracelets, earrings, and pots and pans." The second, *Sooner*, ran out of fuel and ditched in a fiord thirty-five miles southwest of the base. The plane sank in 1,500 feet of water; the entire crew survived. These stories are told in Dorsey, "B-17 Frozen in Ice and Time for 53 Years May Fly Again."
21. A discussion of the details of the weather war can be found in Michael Gjerstad and James Rogers, "Knowledge Is Power: Greenland, Great Powers, and Lessons from the Second World War," Arctic Institute—Center for Circumpolar Security Studies, June 15, 2021. On the German side, a U-boat dropped an unmanned, radio-linked station in far northern Labrador. Known as Station Kurt, it failed after a month and sat undetected until 1977, when a geologist found it while mapping rocks in this remote region of Canada. Details are presented by Michela Rosano, "Throwback Thursday: Nazi Weather Station in Labrador," *Canadian Geographic*, September 9, 2015.
22. This claim of misinformation is stated in a section titled "Nazi Weather Station Suspected" in *The Coast Guard at War*, vol. 2, *Greenland Patrol* (Historical Section, Public Information Division, U.S. Coast Guard Headquarters, 1945), 166.
23. The U.S. Coast Guard captured Nazis in northeastern Greenland. A combat bulletin film, *Nazis Driven from Greenland Coast*, Combat Bulletin no. 38 (War Department, Army Pictorial Service, 1944), shows images of the ship, men with rifles, and the Nazi prisoners. All of this happened during the short summer and became what the film claimed was the northernmost battle of World War II. Most of the captured Nazis were weathermen.
24. "Greenland Inland Ice Weather Stations."
25. The information in this paragraph comes largely from F. Alton Wade, "Wartime Investigation of the Greenland Icecap and Its Possibilities," *Geographical Review* 36, no. 3 (1946): 452–73, https://doi.org/10.2307/210828; and Wallace R. Hansen, *Greenland's Icy Fury*, Texas A&M University Military History Series 32 (College Station: Texas A&M University Press, 1994).
26. The plot of *Night without End* involves a commercial plane that crashes on the ice sheet with its pilot murdered, and a rescue by men in a remote

station on the ice sheet. The station as described is similar to the East Greenland Ice Cap Detachment station, including the Weasels for transport over the snow.

27. Donald M. Taub, "The Greenland Ice Cap Rescue of B-17 PN9E, November 5, 1942, to May 8, 1943" (Coast Guard History Program, n.d.).
28. There are numerous accounts of the winter of 1942 and 1943 on the ice sheet. The most detailed is Taub, "The Greenland Ice Cap Rescue of B-17 PN9E"; the most readable is Stephan Wilkinson, "A Forced Landing on Greenland's Ice Cap Set in Motion One of the Most Extensive—and Costly—Search-and-Rescue Operations Ever Mounted," HistoryNet, September 24, 2020. The entire episode is described in Mitchell Zuckoff, *Frozen in Time: An Epic Story of Survival and a Modern Quest for Lost Heroes of World War II* (New York: Harper, 2014). Zuckoff also describes the first attempt at recovering the Duck and its pilot, radioman, and passenger from the ice in which it's now entombed. That attempt failed.
29. Wesley Craven and James Cate, *The Army Air Forces in World War II*, vol. 4, *The Pacific: Guadalcanal to Saipan* (Office of Air Force History, 1983).
30. H. H. Oliphant, "One of the Toughest Details in the Army," *Yank*, October 22, 1943. This article, in a weekly magazine circulated to millions of U.S. Army soldiers, featured descriptions, stories, and photographs from the Ice Cap Detachment in East Greenland.
31. Oliphant, "One of the Toughest Details in the Army."
32. A thorough history of U.S. and Soviet military involvement in the Arctic is provided by Ronald E. Doel et al., "Strategic Arctic Science: National Interests in Building Natural Knowledge—Interwar Era through the Cold War," *Journal of Historical Geography* 44 (April 2014): 60–80, https://doi.org/10.1016/j.jhg.2013.12.004.
33. Severson, an Air Force pilot, describes bringing the first jet fighters to Thule, including flying in deep cloud cover through narrow fiords and some of Greenland's most awful weather. Eldon Severson, "Frosty Frontier," *Flying Safety* 10 (November 1954): 24–25.
34. Thule meteorologists give a detailed description of the spring storm that led to the record wind gusts in Staff Report, 821st Air Base Group, and 557th Weather Wing, "Two of Thule's Extreme Storms," November 9, 2006.
35. Allen P. Nutman et al., "Rapid Emergence of Life Shown by Discovery of 3,700-Million-Year-Old Microbial Structures," *Nature* 537, no. 7621 (September 22, 2016): 535–38, https://doi.org/10.1038/nature19355.
36. S. J. Mojzsis et al., "Evidence for Life on Earth before 3,800 Million Years Ago," *Nature* 384, no. 6604 (November 7, 1996): 55–59, https://doi.org/10.1038/384055a0.
37. H. C. Larsen et al., "Seven Million Years of Glaciation in Greenland," *Science* 264, no. 5161 (May 13, 1994): 952–55, https://doi.org/10.1126/science.264.5161.952.
38. Ewen Callaway, "Seven-Million-Year-Old Femur Suggests Ancient Human Relative Walked Upright," *Nature*, August 24, 2022: d41586-022-02313–17, https://doi.org/10.1038/d41586-022-02313-7.

39. Henrik Nøhr-Hansen et al., "The Cretaceous Succession of Northeast Baffin Bay: Stratigraphy, Sedimentology and Petroleum Potential," *Marine and Petroleum Geology* 133 (November 2021): 105108, https://doi.org/10.1016/j.marpetgeo.2021.105108.
40. Knutz's story of the coming and going of Greenland's ice sheet is told in Paul C. Knutz et al., "Eleven Phases of Greenland Ice Sheet Shelf-Edge Advance over the Past 2.7 Million Years," *Nature Geoscience* 12, no. 5 (May 2019): 361–68, https://doi.org/10.1038/s41561-019-0340-8.
41. Andrew J. Christ et al., "The Northwestern Greenland Ice Sheet during the Early Pleistocene Was Similar to Today," *Geophysical Research Letters* 47, no. 1 (January 16, 2020), https://doi.org/10.1029/2019GL085176.
42. Bethan Davies, "Calculating Glacier Ice Volumes and Sea Level Equivalents," *Antarctic Glaciers*, September 6, 2023.
43. Scott A. Kulp and Benjamin H. Strauss, "New Elevation Data Triple Estimates of Global Vulnerability to Sea-Level Rise and Coastal Flooding," *Nature Communications* 10, no. 1 (October 29, 2019): 4844, https://doi.org/10.1038/s41467-019-12808-z.
44. Richard B. Alley, *The Two-Mile Time Machine: Ice Cores, Abrupt Climate Change, and Our Future*, Princeton Science Library (Princeton: Princeton University Press, 2014).
45. Guy J. G. Paxman, Jacqueline Austermann, and Kirsty J. Tinto, "A Fault-Bounded Palaeo-Lake Basin Preserved beneath the Greenland Ice Sheet," *Earth and Planetary Science Letters* 553 (January 2021): 116647, https://doi.org/10.1016/j.epsl.2020.116647.
46. Kurt H. Kjær et al., "A Large Impact Crater beneath Hiawatha Glacier in Northwest Greenland," *Science Advances* 4, no. 11 (November 2, 2018): eaar8173, https://doi.org/10.1126/sciadv.aar8173.
47. Gavin G. Kenny et al., "A Late Paleocene Age for Greenland's Hiawatha Impact Structure," *Science Advances* 8, no. 10 (March 11, 2022): eabm2434, https://doi.org/10.1126/sciadv.abm2434.
48. Model results from Clay R. Tabor et al., "Causes and Climatic Consequences of the Impact Winter at the Cretaceous-Paleogene Boundary," *Geophysical Research Letters* 47, no. 3 (February 16, 2020), https://doi.org/10.1029/2019GL085572.
49. Joanna D. Millstein, Brent M. Minchew, and Samuel S. Pegler, "Ice Viscosity Is More Sensitive to Stress than Commonly Assumed," *Communications Earth & Environment* 3, no. 1 (March 10, 2022): 57, https://doi.org/10.1038/s43247-022-00385-x.
50. An easy-to-read explanation of ghost glaciers can be found in Becky Oskin, "'Ghost Glaciers' Protect Greenland's Ancient Landscapes," NBC News, August 6, 2013.
51. Simon L. Pendleton et al., "Rapidly Receding Arctic Canada Glaciers Revealing Landscapes Continuously Ice-Covered for More than 40,000 Years," *Nature Communications* 10, no. 1 (January 25, 2019): 445, https://doi.org/10.1038/s41467-019-08307-w.
52. The basal state of the ice sheet is illustrated and explained by an easy-to-understand NASA web page, "Melt at the Base of the Greenland

Ice Sheet," NASA Earth Observatory, August 4, 2016, and in a more technical manner by Joseph A. MacGregor et al., "A Synthesis of the Basal Thermal State of the Greenland Ice Sheet," *Journal of Geophysical Research: Earth Surface* 121, no. 7 (July 2016): 1328–50, https://doi.org/10.1002/2015JF003803.

53. MacGregor et al., "A Synthesis of the Basal Thermal State of the Greenland Ice Sheet."
54. Michiel Van Den Broeke et al., "Partitioning Recent Greenland Mass Loss," *Science* 326, no. 5955 (November 13, 2009): 984–86, https://doi.org/10.1126/science.1178176.
55. Matthew Cooper and Laurence Smith, "Satellite Remote Sensing of the Greenland Ice Sheet Ablation Zone: A Review," *Remote Sensing* 11, no. 20 (October 16, 2019): 2405, https://doi.org/10.3390/rs11202405.
56. J. C. Ryan et al., "Greenland Ice Sheet Surface Melt Amplified by Snowline Migration and Bare Ice Exposure," *Science Advances* 5, no. 3 (March 2019): eaav3738, https://doi.org/10.1126/sciadv.aav3738.

CHAPTER 2: FROZEN WATER

1. The Smithsonian provides an overview of Bentley's life and an extensive bibliography of his work in "Wilson A. Bentley: Pioneering Photographer of Snowflakes" (web page), Smithsonian Institution Archives, September 14, 2012.
2. The Buffalo Museum of Science holds many of Bentley's original negatives. In the museum's magazine, *Hobbies*, Dr. Carlos Cummings wrote a four-page tribute to Bentley in 1947 when the museum acquired the collection of nearly 12,000 glass plates: Carlos E. Cummings, "Wilson Alwyn Bentley Snowflake Plates Acquired," *Hobbies* 28, no. 2 (1947): 36–39.
3. U.S. Customs records show that the Baders (listed as Swiss citizens), four other Europeans, and seven Americans flew north from the Caribbean to Miami with stops in Aruba, Haiti, and Cuba on May 31, 1945. The plane was a KLM Lockheed Lodestar, a two-engine tail-dragger that held eighteen passengers. It had tail letters PJ-AKA. The plane entered service in 1943 and was scuttled in 1948, dumped off a cliff at the Curaçao airport into the ocean. The history of this aircraft is given by Arno Landewers, "The Complete Aircraft Register of the Dutch Antilles" (website), July 10, 2023.
4. In 1949, Henri Bader's declaration of intention to become a U.S. citizen indicates that the couple lived right across the street from the University, at 12 Suydam Street in New Brunswick, New Jersey. Because of his years in South America, where the declaration states he married Adele on 9/24/1938 in Santiago del Estro, Argentina, Bader spoke fluent Spanish. Today, Suydam Street is a Latin neighborhood with a Venezuelan bakery, a Dominican restaurant, and several Mexican groceries.
5. An Argentine government website ("Inicio," *CEMLA—Centro de Estudios Migratorios Latinoamericano*) provides immigration data for people

coming to the country. It shows that Bader, age 31, single, and a Swiss citizen, departed in July from Amberes (Antwerp) on a vessel of Belgian registry, the *Piriapolis,* and arrived on 8/23/1938. His occupation is listed as Mineralogista (Mineralogist).

6. In less than two years, the *Piriapolis* sailed eight times to Buenos Aires. Its last arrival was on March 10, 1940, according to "Ship PIRIAPOLIS—Arrivals to Argentina," Hebrewsurnames.com, which maintains a list of ships and their passengers to Argentina.
7. Details of how the tournament evolved with many of the players' countries at war are told by Andrew Smith, "The Infamous 8th Chess Olympiad," *ThePawnSlayer* (blog), chess.com, September 4, 2020.
8. I found no Argentine records of Adele arriving by ship, although there are emigration papers indicating she left Switzerland in October 1922 bound for Buenos Aires on the Steamship *Gelria* from Cherbourg, France. She would have just turned 18, as U.S. Social Security records list her birthday as July 3, 1904. Reuben Goossens, "S.S. *Gelria* Completed in 1913 of the Royal Holland Lloyd," ssmaritime.com, provides a detailed history of the vessel and suggests that "in addition to all the grandiose luxury of First and Second Class, in Steerage there would always be a variety of migrants looking for a new life, such as many refugees who were escaping Eastern Europe, such as Polish farmers and other poor, as well as Jews who suffered extreme anti-Semitism." Immigration documents such as those for the Bader's return to New York on the *Queen Elizabeth* provide information about the couple. Adele was 5'2" with Argentine relatives, green eyes, and brown hair. Henri was 5'7". They were born ten miles apart, in Lenzburg (Adele) and Brugg (Henri), Switzerland. Adele was two and a half years older than Henri.
9. Marcel de Quervain and Hans Rothlisberger, "Obituary—Henri Bader (1907–1998)," *Ice: News Bulletin of the International Glaciological Society,* no. 120 (1999). The obituary begins with this story: "Henri Bader died in a hospital in Miami, Florida, on 6 December 1998, at the age of nearly 92. He was extremely ill with emphysema and had a virulent intestinal infection. Visiting him in March 1997, one of us found him frail and forgetful, but still vividly interested in what was going on in glacier research. He also proudly placed our order in forceful Spanish when we joined him at his favorite restaurant. He told us that one had to master Spanish to get around on Miami streets."
10. Several catalogues provide details of the wreck. The most detail is provided by "MV *Piriapolis* [+1940]," Wrecksite, February 4, 2008, https://wrecksite.eu/wreck.aspx?15062. The *Piriapolis* is now a dive site with the wreck broken into several pieces on a sandy and shell-rich seabed.
11. Much of the information about the Baders during and after World War II comes from two articles in *Ice,* a magazine published several times a year by the Glaciological Society, one a 1964 description of his life and career: "Henri Bader," *Ice: News Bulletin of the Glaciological Society,* no. 13 (December 1963), and the other his obituary: de Quervain and Rothlisberger, "Obituary—Henri Bader (1907–1998)."

12. The mine contained two phosphorus minerals, apatite and whitlockite, both a mix of calcium, phosphorus, and oxygen, like teeth and bones. Part 2, "Phosphate Rock and Superphosphate," in *World Trade in Commodities* (Washington, DC: U.S. Bureau of Foreign and Domestic Commerce, 1947).
13. The Curaçao refineries processed Venezuelan crude. Gasoline went from Curaçao and Aruba west to the Pacific Theatre through the Panama Canal, and east to Europe and North Africa across the Atlantic Ocean. In early 1942, a German U-boat lobbed shells at Curaçao's shore but quickly submerged when American forces returned fire. Then, later that year, another U-boat torpedoed and sank an American gunboat, the U.S.S. *Erie*, on December 5 in the harbor at Willemstad, Curaçao.
14. See memo on page 24, section 3 of University of Minnesota Institute of Technology Engineering Experiment Station, *Interim Report to Snow, Ice, and Permafrost Research Establishment* (Wilmette, IL: Snow, Ice, and Permafrost Research Establishment, Corps of Engineers, U.S. Army, 1950), for details of what Bader did in Europe.
15. Henri Bader, "Trends in Glaciology in Europe," *Geological Society of America Bulletin* 60, no. 9 (1949): 1309, https://doi.org/10.1130/0016-7606(1949)60[1309:TIGIE]2.0.CO;2.
16. University of Minnesota Institute of Technology Engineering Experiment Station, *Interim Report to Snow, Ice, and Permafrost Research Establishment.* This early report from SIPRE lays out, in limited text and numerous flow charts, the organization of snow and ice studies that SIPRE would conduct over the next decade. The U.S. Army, and scientists working with the Army, accomplished most of what the report suggested. Much of that work was done in Greenland, including studies of snow, ice, and permafrost properties, change over time, and excavation methods. The report presents a basic science plan with the requisite justification for military needs.
17. A description of the U.S. military's involvement in the Juneau Icefield Research Project (JIRP) and the mutual benefits of scientists and the military working together, as well as the logistics of air support on the ice field, are in Maynard M. Miller, "Adventures in Logistics," *Mountaineer,* December 15, 1951.
18. Details of science and work are in Maynard M. Miller, "Instruments and Methods: Englacial Investigations Related to Core Drilling on the Upper Taku Glacier, Alaska," *Journal of Glaciology* 1, no. 10 (January 1951): 579–80, https://doi.org/10.3189/S0022143000026320.
19. Nearby Camp 10 is still in use today by JIRP. Camp 10B was the temporary drilling camp on the ice.
20. The story of JIRP is well told by Scott Yorko, "A 70-Year-Old Glacial Study Almost Died When the World Needed It Most," *Popular Mechanics,* September 12, 2018.
21. Maynard M. Miller, "The Vanishing Glaciers," *Science Illustrated,* March 1949.
22. All of these details come from an oral history interview of Wasserburg on April 25, May 3, May 10, and May 17, 1995, which is available online

as Gerald J. Wasserburg, oral history interview with David Valone, 2017, CaltechOralHistories, Caltech Institute Archives.

23. Wasserburg wrote his own autobiography as a peer-reviewed paper: G. J. Wasserburg, "Isotopic Adventures—Geological, Planetological, and Cosmic," *Annual Review of Earth and Planetary Sciences* 31, no. 1 (May 2003): 1–74, https://doi.org/10.1146/annurev.earth.31.100901.141409.
24. Miller, "Instruments and Methods," presents a modest history of this work, including photographs of the camp and drilling rig. It was published in the first volume of the *Journal of Glaciology*, which would go on to publish many papers relevant to ice cores, ice coring, and climate.
25. An exceptionally detailed history of ice-drilling technologies over more than 200 years, including Agassiz's early work, is presented in Pavel G. Talalay, *Mechanical Ice Drilling Technology* (Singapore: Springer Singapore, 2016), https://doi.org/10.1007/978-981-10-0560-2.
26. Camille Hankel et al., "Louis Agassiz," Department of Earth and Planetary Sciences, Harvard University (website), 2023, discusses the evidence of both plagiarism and racism.
27. Robert Krulwich, "Strange-Looking Tombstone Tells of Moving Ice, Ancient Climates and a Restless Mind," December 5, 2012, in *Krulwich Wonders*, National Public Radio.
28. Films (16 mm) taken of early JIRP expeditions document the drilling. Reel 94 has extensive footage of the team tripping rods in and out of the borehole and some footage of the ice core along with the water pump and what looks like Wasserburg having fun with another team member. Reel 91 shows the ice core, ice coring, and Wasserburg entering the science trench used for core processing. Digitized versions are available online from Juneau Icefield Research Project, AGSL Digital Photo Archive—North and Central America, University of Wisconsin–Milwaukee Libraries, 1950.
29. Chester Langway, "Stratigraphic Analysis of a Deep Ice Core from Greenland," vol. 125, Geological Society of America Special Papers (Boulder: Geological Society of America, 1970), https://doi.org/10.1130/SPE125, describes these details in the section "Early deep drilling."
30. Henri Bader, "Introduction to Ice Petrofabrics," *Journal of Geology* 59, no. 6 (1951): 519–36.
31. Miller, "Instruments and Methods."
32. Bader, "Trends in Glaciology in Europe."
33. Tim Lydon, "The Mighty Taku Glacier Takes a Bow," *Hakai*, June 28, 2021.
34. Lydon, "The Mighty Taku Glacier Takes a Bow."
35. Talalay, *Mechanical Ice Drilling Technology*.
36. This makes sense, as Chicago was at the time the hub of American isotope research. Harold Urey and Sam Epstein were there as researchers, and Wasserburg was a student there at the time.
37. Wasserburg's last visit was in 1995, three years before Henri died. In his oral history interview with David Valone, Wasserburg said of the Baders, "I try to visit with him and his wife to this day, as much as I can.

He's now quite old and feeble. I visited him last December, and they're in bad shape, mostly just due to age. No children, and they both basically adopted me as their child in an intellectual sense." Faculty at the University of Miami told me that Wasserburg donated funds to support the Annual Henri and Adele Bader Lectureship at the school. Lonnie Thompson was one of the invited lecturers.

38. A review of these projects and the aircraft that serviced them is presented by Stephen Mock, "Greenland Operations of the 17th Tactical Airlift Squadron and CRREL," The Firebird Association (website), March 1973, http://www.firebirds.org/menu1/mnu1_p12.htm.
39. The tangle of organizations is explained in Arctic Construction and Frost Effects Laboratory, *Project Mint Julep: Investigation of a Smooth Ice Area of the Greenland Ice Cap*, Report (U.S. Army Corps of Engineers, June 1954): "In the summer of 1953, the Arctic Construction and Frost Effects Laboratory participated in Project Mint Julep, an investigation of the Greenland Ice Cap led by the American Geographical Society under contract with the Air University, U.S. Air Force, for the purpose of continuing investigations of the construction and maintenance of airfields on the Ice Cap begun in 1947 on Project Snowman."
40. F. Alton Wade, "Wartime Investigation of the Greenland Icecap and Its Possibilities," *Geographical Review* 36, no. 3 (1946): 452–73, https://doi.org/10.2307/210828.
41. Arctic Construction and Frost Effects Laboratory, *Final Report on Preparations of Frost Effects Laboratory for Project Overheat* (Boston: Army Corps of Engineers, 1950).
42. In 1947, Air Force pilots from Alaska, returning from a mission across the pole touching the edge of Russian air space as the Cold War began to heat up, lost their way. After flying for hours with the stars too dim, and the sun refusing to get far enough above the horizon, for sextant navigation, they ran out of fuel. To survive, they set down their two-year-old B-29 Superfortress (nicknamed the *Keebird*) on a frozen lake in what turned out to be northwestern Greenland, not Alaska. They were way off course, having landed 200 plus miles northeast of Site 1 and 300 plus miles from Thule Air Base. A rescue plane managed the risky landing on snow-covered ice and picked up the plane's crew. True Comics, an educational comic book series whose motto was "Truth is stranger than fiction," picked up on the *Keebird* story. The August 1947 edition features a three-page spread detailing the crash and the rescue. It's titled "Rocket Rescue," in honor of the JATOs used for the rescue. The *Keebird* sat frozen and preserved for almost a half century before an unusual salvage attempt in 1995. The goal was to fly the fifty-year-old plane off the ice. Given four new engines and four new propellors as it sat on the frozen lake, the *Keebird* taxied across the ice. But a fire, caused by a failed generator fuel line, sent the unique plane back to its icy grave before it could ever take off.
43. Arctic Construction and Frost Effects Laboratory, *Project Mint Julep*.
44. Edward R. Jackovich and Albert F. Wuori, *Performance Testing of a Snow-*

blast Plow, Special Report (Hanover, NH: Cold Regions Research and Engineering Laboratory, 1966).

45. Arctic Construction and Frost Effects Laboratory, *Project Mint Julep*. The report has one cross section, and it shows drilling logs for eight of the thirty-four holes.
46. Arctic Construction and Frost Effects Laboratory, *Project Mint Julep*.

CHAPTER 3: INTO THE ICE

1. F. Salzar, "Claiming Ultima Thule," *Hakai*, September 8, 2020.
2. Pytheas's Thule might have been Iceland or Greenland, although the latest geographical analysis suggests it was most likely an island off the west coast of Norway. See Cameron McPhail, "Pytheas of Massalia's Route of Travel," *Phoenix* 68, no. 3/4 (2014): 247, https://doi.org/10.7834/phoenix.68.3-4.0247.
3. Jon Gertner, *Ice at the End of the World: An Epic Journey into Greenland's Buried Past and Our Perilous Future* (New York: Random House, 2019), provides an excellent history of Thule from the first explorers through today.
4. David Arnold, "Minnesotans at the Top of the World," *Minnesota History*, Spring 2016.
5. By the end of 1952, the runway had been expanded to 10,000 feet, according to Kevin L. Bjella, *Thule Air Base Airfield White Painting and Permafrost Investigation: Phases I–IV*, TR-ERDC/CRREL 13-8 (Hanover, NH: Cold Regions Research and Engineering Laboratory, 2013).
6. Arnold, "Minnesotans at the Top of the World."
7. DeNeen L. Brown, "Trail of Frozen Tears," *Washington Post*, October 22, 2002.
8. Agence France-Presse, "The Arctic: Greenland: Inuit Lose A 50-Year Court Battle," World Briefing, *New York Times*, November 29, 2003.
9. Sandra Erwin, "Space Force Renames Greenland's Thule Air Base," *SpaceNews*, April 6, 2023.
10. The U.S. Army produced a series of television shows under the name *The Big Picture*. Over two decades, the Army sent episodes gratis to TV stations around the United States, which broadcast many of them. *Operation Blue Jay* aired as episode #227. Most episodes are available today on YouTube.
11. The movie starts with a slow pan of a world map, focusing on the American air and missile defense system, first along the Canadian border, then in central Canada, and finally, the Distant Early Warning radar line across northern Canada and Greenland. The first ten minutes of the film feel like a military newsreel because, for the most part, they are. There are images of snowy Arctic camps (Camp Tuto) and footage of the construction of Thule Air Base, then code-named Operation Blue Jay. The main base in *The Deadly Mantis* was a different bird of a different color, Red Eagle One.
12. See a wide-ranging analysis of the big bug movie genre, and a review of

critical interpretations of these movies in a political and social context, in William M. Tsutsui, "Looking Straight at *Them!* Understanding the Big Bug Movies of the 1950s," *Environmental History* 12, no. 2 (April 1, 2007): 237–53, https://doi.org/10.1093/envhis/12.2.237.3.

13. "Greenland Inland Ice Weather Stations," in *Encyclopedia Arctica*, vol. 7: *Meteorology and Oceanography* (Digital Collections, Dartmouth College Library).
14. Most of the information in this discussion comes from Paul-Emile Victor, "Wringing Secrets from Greenland's Icecap: In Six Years of Exploration, Daring and Resourceful Frenchmen Map a Lost World Buried under Thousands of Feet of Ice," *National Geographic*, January 1, 1956.
15. Weather records there, collected by the men, extend from September 1949 through August 1951. See George Weidner and Charles Stearns, "A Two-Year Record of the Climate on the Greenland Crest from an Automatic Weather Station" (Wisconsin University at Madison, Dept. of Meteorology, March 1992).
16. Clifford Hicks, "Coldest Life on Earth," *Popular Mechanics*, November 1951.
17. Jon Gertner tells the story of the American visit to the camp (based on his interview with Carl Benson) in great detail in *Ice at the End of the World*.
18. Victor, "Wringing Secrets from Greenland's Icecap."
19. The quotations in this paragraph come from Victor, "Wringing Secrets from Greenland's Icecap."
20. Many newspapers published articles about Sites 1 and 2. For example, the *Des Moines Register*, December 21, 1954, featured the camps on page 1. Papers at the time described these bases as weather stations designed for "year 'round living" and staffed by Air Force personnel. This description fits with weather records available from both sites between 1953 and 1956. The photographs in the *Register* (and the one reproduced in this book) must be Site 1 because the horizon is not flat, but rather shows rocky hills. Site 2 was too far up ice (nearly 200 miles from the ice margin) for there to be any rocky outcrops.
21. B. Lyle Hansen, *Instrumentation of Ice Cap Stations* (Wilmette, IL: Snow, Ice, and Permafrost Research Establishment, Corps of Engineers, U.S. Army, April 1955).
22. James J. Haggerty Jr., "Subways under the Icecap," *Collier's*, May 11, 1956.
23. Photographs taken by Gene P. McManus at the camps show these items and events. They are available in our online image archive, Iceworm: Camp Century and the U.S. Military/Science Complex, developed by the University of Vermont under support from the U.S. National Science Foundation, at https://campcentury.omeka.net/, and we obtained them from Radomes, Inc., the website of the Air Defense Radar Veterans' Association.
24. The film, *Icecap Airmen Live Underground*, AFNR #8, 1956, starts with the men climbing out of hatches, the only part of the camp visible on the surface. They shovel snow into snow melters, and a delivery of food

and supplies, including barrels of fuel, comes in on a C-47 ski-equipped plane.

25. A very readable account of flying C-119s is provided in Jim Blackburn, "Arctic Adventures," May 2015, available online as part of an extensive history of the C-119 at https://www.ruudleeuw.com/pdf/c119-blackburn_arctic_adventures.pdf.
26. Malcolm Mellor, *Methods of Building on Permanent Snowfields*, Report (Hanover, NH: Cold Regions Research and Engineering Laboratory, October 1968).
27. Richard F. Barquist and Chimer D. Moore Jr., *Report, Medical Section, of a Greenland Operation, Summer 1957* (Fort Knox, KY: U.S. Army Medical Research Laboratory, 1957).
28. Thomas R. Ostrom, Charles R. West, and James J. Shafer, "Investigation of a Sewage Sump on the Greenland Icecap," *Journal (Water Pollution Control Federation)* 34, no. 1 (1962): 56–62.
29. Henri Bader and F. Small, *Sewage Disposal at Ice Cap Installations*, SIPRE Report (Wilmette, IL: Snow, Ice, and Permafrost Research Establishment, Corps of Engineers, U.S. Army, 1955).
30. Malcolm Mellor, *Utilities on Permanent Snowfields*, Report (Hanover, NH: Cold Regions Research and Engineering Laboratory, October 1969).
31. Ostrom, West, and Shafer, "Investigation of a Sewage Sump on the Greenland Icecap."
32. Buss discusses the effect on soldiers in the Arctic with particular reference to Thule and Sites 1 and 2 in Lydus H. Buss, *US Air Defense in the Northeast 1940–1957*, Historical Reference Paper Number One (Directorate of Command History, Office of Information Services, Headquarters Continental Air Defense Command, 1957).
33. All of the information in the above paragraph, including the quotation, is drawn from Eldon Severson, *The History of the 64th Air Division, 1 Jan–30 June 1954* (Office of Information Services, 1954), as part of 1,700 pages of microfilm transferred to PDF from originals at Maxwell Air Force Base, reel PO639, filmed 10/10/1972.
34. Severson, *History of the 64th Air Division.* Dessert recipes stretched over 58 pages and included Foamy Chocolate Cake with Yeast, Chocolate Cake with Sour Milk, Marble Loaf Cake with Evaporated Milk, Lucky Nut Loaf, Press Cookies ("nice for teas and special occasions"), and Coconut Meringue Kisses.
35. Carl S. Benson and Richard H. Ragle, *Project Jello: SIPRE Greenland Expedition, 1955*, Report (Wilmette, IL: U.S. Army Snow, Ice, and Permafrost Research Establishment, February 1956).
36. Site 1 and Site 2 had many aliases. Site 1 was also Etah Air Station, G-33, N-33, and the A-site. Site 2 was also Ice Cap Air Station, G-34, N-34, and the B-site. Every report seemed to use different names; perhaps this was by intent, as the sites, at the time of construction and for several years after, were secret.
37. For an example, see text and a photograph with extended caption in

"Army Gets Mole Type of Arctic Tent," *Austin American*, December 21, 1954.

38. Bruce Jacobs, "Our Secret Bases under the Polar Ice Cap," *Real—the Exciting Magazine for Men*, February 1955.
39. All of the quotes in the above paragraph come from Jacobs, "Our Secret Bases."
40. Severson, *History of the 64th Air Division.*
41. The radar units at Sites 1 and 2 were portable—unlike the massive Distant Early Warning (DEW) line or Ballistic Missile Early Warning System (BMEWS) positioned elsewhere in the Arctic. These modular radar units came in a set of boxes that four airmen could carry. The plane-finding radars were good to distances of 160 miles. The fabric-covered dome visible in the Air Force newsreel *Icecap Airmen Live Underground* is the weatherproof radome that sheltered the radar antenna.
42. A decade after the Army installed the tubes, they had not collapsed, despite more than 20 feet of snow fully burying them. Extensive measurements of their deformation, which is remarkably slight, are presented in Malcolm Mellor, *Undersnow Structures: N-34 Radar Station, Greenland* (Hanover, NH: U.S. Army Materiel Command, Cold Regions Research and Engineering Laboratory, 1964).
43. John Able Jr., "Ice Tunnel Closure Phenomena" (master's thesis, Colorado School of Mines, 1959).
44. Donald O. Rausch, *Ice Tunnel, TUTO Area, Greenland, 1956*, Technical Report 44 (Wilmette, IL: U.S. Army Snow, Ice, and Permafrost Research Establishment, Corps of Engineers, 1958).
45. Rausch, *Ice Tunnel, TUTO Area, Greenland, 1956.*
46. Haggerty, "Subways under the Icecap."
47. A map that shows tunnel dimensions was published in George K. Swinzow, "Investigation of Shear Zones in the Ice Sheet Margin, Thule Area, Greenland," *Journal of Glaciology* 4, no. 32 (1962): 215–29, https://doi.org/10.3189/S0022143000027416.
48. Frank Russell, *An Under-Ice Camp in the Arctic*, Special Report (Hanover, NH: U.S. Army Materiel Command, Cold Regions Research and Engineering Laboratory, 1961).
49. The details of the early tunnels are provided in Theodore R. Butkovich, *Some Physical Properties of Ice from the TUTO Tunnel and Ramp, Thule, Greenland*, Report (Wilmette, IL: U.S. Army Snow, Ice, and Permafrost Research Establishment, May 1959).
50. Data from numerous surveys using a variety of techniques over several years all show rapid tunnel closure. John F. Abel, *Ice Tunnel Closure Phenomena*, Report (Wilmette, IL: U.S. Army Snow, Ice, and Permafrost Research Establishment, January 1961).
51. Data from these experiments are in Butkovich, *Some Physical Properties of Ice from the TUTO Tunnel and Ramp.*
52. The story is told in rather dramatic fashion by Walter Sullivan, "Army Ice Tunnel Inaugurated by Sheltering 25 in a Blizzard," *New York Times*, October 7, 1961.

53. Frank L. Russell, *Water Production in a Polar Ice Cap by Utilization of Waste Engine Heat* (Hanover, NH: U.S. Army Cold Regions Research and Engineering Laboratory, 1965).
54. George K. Swinzow, *Preliminary Investigations of Permacrete,* Report (Hanover, NH: Cold Regions Research and Engineering Laboratory, February 1965).
55. John F. Abel, *Permafrost Tunnel, Camp TUTO, Greenland,* Report (Wilmette, IL: U.S. Army Snow, Ice, and Permafrost Research Establishment, October 1960).
56. Details of this excavation and how it deformed are provided in Henri Bader et al., *Excavations and Installations at SIPRE Test Site, Site 2, Greenland,* Report (Wilmette, IL: U.S. Army Snow, Ice, and Permafrost Research Establishment, April 1955).
57. Bader et al., *Excavations and Installations at SIPRE Test Site.*
58. Mellor, *Methods of Building on Permanent Snowfields.*
59. See footnote on page 8 of Elmer F. Clark, *Camp Century: Evolution of Concept and History of Design Construction and Performance,* CRREL Technical Report 174 (Hanover, NH: Cold Regions Research and Engineering Laboratory, 1965).
60. Clark, *Camp Century,* reviews the history of below-snow camps, and construction techniques used to create them, in great detail.
61. Alfred Fuchs, *Structure of Age-Hardening Disaggregated Peter Snow,* Research Report 53 (Wilmette, IL: U.S. Army Snow, Ice, and Permafrost Research Establishment, Corps of Engineers, 1960).
62. Gunars Abele, *Snow Roads and Runways,* CRREL Monograph (Hanover, NH: U.S. Army Cold Regions Research and Engineering Laboratory, 1990).
63. A well-illustrated and extensive summary of the Army's work making camps in ice sheets, including the first trenches cut by Peter plow at Site 2, can be found in Mellor, *Methods of Building on Permanent Snowfields.*
64. Barquist and Moore, *Report, Medical Section, of a Greenland Operation, Summer 1957.*
65. This quotation, and the information in the paragraph above it, is from Bledsoe's obituary, James A. Bender, "Lucybelle Bledsoe—1923–1966," *Journal of Glaciology* 6, no. 47 (1967): 755–56, https://doi.org/10.3189/S0022143000020001.
66. The risk was real, but the often-told story of Eisenhower demanding that all lesbians be removed from the military serving underneath him in Europe appears not to be true: see Donna Knaff, "The 'Ferret Out the Lesbians' Legend: Johnnie Phelps, General Eisenhower, and the Power and Politics of Myth," *Journal of Lesbian Studies* 13, no. 4 (October 27, 2009): 415–30, https://doi.org/10.1080/10894160903048155. As president, Eisenhower issued an executive order related to national security that included the wording "(iii) Any criminal, infamous, dishonest, immoral, or notoriously disgraceful conduct, habitual use of intoxicants to excess, drug addiction, sexual perversion. . . ." In the 1950s and 1960s, the U.S. government considered homosexuality a

perversion and thus grounds for rejection or dismissal from government service.

67. Lucy Bledsoe discusses this in detail in the postscript of Lucy Jane Bledsoe, *A Thin Bright Line* (Madison: University of Wisconsin Press, 2016).
68. Information about Langway's career and life history comes from "A Chilling Conversation with UB's Ice Core King," *Buffalo News*, April 12, 1989; the 1956 Boston University yearbook; and communication with the Boston University library. He received an A.M. degree, and it appears he did not draft a thesis for that degree, but only took classes.
69. Kim A. McDonald, "N.Y. Scientist Uncovers Clues to Earth's Ancient Climate within the Frozen Remnants of the Last Ice Age," *Chronicle of Higher Education*, September 27, 1989.
70. Chester C. Langway, *The History of Early Polar Ice Cores*, ERDC/CRREL Technical Report (Hanover, NH: U.S. Army Corps of Engineers, Engineer Research and Development Center, Cold Regions Research and Engineering Laboratory, 2008) reviews the drilling of the early American ice cores.
71. Chester Langway, "Stratigraphic Analysis of a Deep Ice Core from Greenland," vol. 125, Geological Society of America Special Papers (Boulder: Geological Society of America, 1970), https://doi.org/10.1130/SPE125.
72. Chester C. Langway, "A 400 Meter Deep Ice Core in Greenland: Preliminary Report," *Journal of Glaciology* 3, no. 23 (1958): 217, https://doi.org/10.3189/S0022143000024278.
73. "1100 AD Icicle Studied by the Army," *Cincinnati Enquirer*, February 8, 1959.
74. "1100 AD Icicle Studied by the Army."
75. M. M. Miller, *Juneau Icefield Research Project, Alaska 1950 Summer Field Season*, JIRP Report (American Geographical Society, Department of Exploration and Field Research, 1954) is a report of the 1950 activities on the ice field. Although written by Miller, it's clear Bader had a strong influence on this document. Almost all the analyses Langway did on the Site 2 core were either proposed for or done on the Taku core many years earlier.
76. Langway, "Stratigraphic Analysis of a Deep Ice Core from Greenland."
77. W. Dansgaard, "The Abundance of O^{18} in Atmospheric Water and Water Vapour," *Tellus* 5, no. 4 (November 1953): 461–69, https://doi.org/10.1111/j.2153-3490.1953.tb01076.x.
78. The isotope data were measured, interpreted, and published in less than two years in Samuel Epstein and Robert Sharp, "Oxygen Isotope Studies," *National Academy of Sciences, IGY Bulletin #21 (Transactions, American Geophysical Union)* 40, no. 1 (March 1959): 81–84. It's perplexing to me that Langway never published the isotope data in a peer-reviewed paper and never published with Epstein; Langway is not a co-author on the Epstein and Sharp paper. In 1965, eight years after the Site 2 core was drilled, Langway finished his doctoral thesis in geology: Charles Langway, "Stratigraphic Analysis of a Deep Ice Core from Greenland" (doc-

toral thesis, University of Michigan, 1965). For a short time, it defined the then nascent field of ice-core analysis. It was a substantial 466 pages. Inside was a full analysis of the 1957 core from Site 2 that Langway had first analyzed in the field. The Army published the same text as a report in 1967: Chester C. Langway, *Stratigraphic Analysis of a Deep Ice Core from Greenland* (Hanover, NH: U.S. Army Materiel Command, Cold Regions Research and Engineering Laboratory, 1967), and so did the Geological Society of America in 1970: Chester Langway, "Stratigraphic Analysis of a Deep Ice Core from Greenland," vol. 125, Geological Society of America Special Papers (Boulder: Geological Society of America, 1970), https://doi.org/10.1130/SPE125. By 1970, the 13-year-old Site 2 data had much less relevance. The Camp Century core, five times longer and reaching ice a hundred times older, was all the rage since the presentation of extensive isotope data by W. Dansgaard et al., "One Thousand Centuries of Climatic Record from Camp Century on the Greenland Ice Sheet," *Science* 166, no. 3903 (October 17, 1969): 377–81, https://doi.org/10.1126/science.166.3903.377.

79. Miller, "Vanishing Glaciers."
80. Henri Bader, "The Problem of Arctic Research," in *Proceedings of the Third Annual Arctic Planning Session, November 1960*, GRD Research Notes 55 (Bedford, MA: Geophysical Research Directorate, 1961), 109–13. This obscure report is available through Google Books.
81. Eunice Foote, "Circumstances Affecting the Heat of Sun's Rays," *American Journal of Art and Science* 22, no. 116 (November 1856): 382–83. Foote's paper was read by Professor Joseph Henry of the Smithsonian in front of the American Association for the Advancement of Science at its meeting held in Albany, New York, on August 23, 1856.
82. An excellent review of the early scientific work on the influence of carbon dioxide on the control of Earth's radiation balance, and thus its climate, is provided by Roland Jackson, "Eunice Foote, John Tyndall and a Question of Priority," *Notes and Records: Royal Society Journal of the History of Science* 74, no. 1 (March 20, 2020): 105–18, https://doi.org/10.1098/rsnr.2018.0066.
83. Gilbert N. Plass, "The Carbon Dioxide Theory of Climatic Change," *Tellus* 8, no. 2 (May 1956): 140–54, https://doi.org/10.1111/j.2153-3490.1956.tb01206.x.
84. Peter M. Cox, Chris Huntingford, and Mark S. Williamson, "Emergent Constraint on Equilibrium Climate Sensitivity from Global Temperature Variability," *Nature* 553, no. 7688 (January 2018): 319–22, https://doi.org/10.1038/nature25450.
85. Spencer Weart, "Roger Revelle's Discovery," The Discovery of Global Warming (website), May 2023, https://history.aip.org/climate/Revelle.htm.
86. You can see the concentrations of carbon dioxide in the atmosphere today and in the past, including Keeling's data, at The Keeling Curve (website), Scripps Institution of Oceanography, https://keelingcurve.ucsd.edu/.

87. Charles D. Keeling, "The Concentration and Isotopic Abundances of Carbon Dioxide in the Atmosphere," *Tellus* 12, no. 2 (May 1960): 200–203, https://doi.org/10.1111/j.2153-3490.1960.tb01300.x.
88. Henri Bader, "The Significance of Air Bubbles in Glacier Ice," *Journal of Glaciology* 1, no. 8 (1950): 443–51, https://doi.org/10.3189/S002214300001279X. Jon Gertner, in *Ice at the End of the World*, reports that Bader was not the first to see bubbles in ice. Men stationed on the west coast of Greenland dug a pit similar to Sorge's, but in ice rather than firn, and saw bubbles in samples of ice from that pit (180–81).
89. Science Service was the brainchild of newspaper magnate E. W. Scripps, the same man who helped fund the Scripps Institute of Oceanography, where Roger Revelle and Charles Keeling worked. There are numerous versions of Segman's syndicated article based on his interview with Bender in newspapers around the country; for example, "Bubbles Trapped, 'Ancient Air' Held in Polar Ice Cores," *Tallahassee Democrat* (Tallahassee, Florida), February 14, 1959, 6.

CHAPTER 4: WARMTH UNDER THE SNOW

1. Herbert O. Johansen, "U.S. Army Builds a Fantastic City under Ice," *Popular Science*, February 1960.
2. George Amick, "Dig Deep for Gracious Living beneath Greenland's Icecap," *Cincinnati Enquirer*, December 13, 1958; "For Survival, Dig In," *Cincinnati Enquirer*, December 21, 1958.
3. Frank J. Leskovitz, "Camp Century, Greenland," Science Leads the Way (website), https://gombessa.tripod.com/scienceleadstheway/id9.html.
4. Elmer F. Clark, *Camp Century: Evolution of Concept and History of Design Construction and Performance*, CRREL Technical Report 174 (Hanover, NH: Cold Regions Research and Engineering Laboratory, 1965).
5. Most of these recollections are from Ray Hansen, who spent time at Camp Century, Tuto, and Thule. His recollections are published on one of numerous Thuleforum web pages at Ray Hansen, "Ray Hansen's Pictures," Thuleforum, April 7, 2012, http://www.thuleforum.dk/ray/. Hansen includes numerous photographs, mostly unavailable elsewhere. He recalls that at least some Army Corps engineers seemed to enjoy these rides, and decades later described the stoves used for warmth and the stories old-timers told on these trips.
6. Here's the calculation: A Big Mac contains 540 kilocalories (kcal) of energy, according to McDonald's; that's equivalent to 2,260 kilojoules (kJ). The latent heat of fusion (melting water) is 334 kJ/kg, and the latent heat of vaporization (making steam from liquid water) is 2,256 kJ/kg. A kilocalorie is equal to 4.185 kJ, and a pound is 0.120 gallons of water. From this, I calculate that the energy in one Big Mac could melt 6.77 kg of ice (14.88 pounds) if extracted with 100 percent efficiency, yielding 1.78 gallons of water.
7. Jon Gertner, *Ice at the End of the World: An Epic Journey into Greenland's*

Buried Past and Our Perilous Future (New York: Random House, 2019), tells this story in detail.

8. Raul Rodriguez, *Development of Glacier Subsurface Water Supply and Sewage Systems*, vol. 1737, U.S. Army Corps of Engineers, Technical Report (Ft. Belvoir: Defense Technical Information Center, 1963).
9. Richard Schmitt and Raul Rodriguez, "Glacier Water Supply System," *Military Engineer* 52, no. 349 (1960): 382–83.
10. The story comes from Dagmar Cole and involves her father, Oliver Gene Drummond, who helped install the nuclear reactor. I provide more details about Drummond later in this chapter. Cole emailed me: "Quick family story. . . . We were like many enlisted families, living on a very tight and limited budget. My mother scrimped and saved up so she could cook a nice meal for my father when he came home. It was after his first visit, so it would have been July 1960. He arrived late in the afternoon, said he was starving, what was for dinner? My mother proudly announced, STEAK! And my father's face fell, and he was obviously so disappointed. My mother asked him what was the matter? He told her the Army tried to keep their calories up, so they ate steak nearly EVERY DAY and he was sick of steak, he didn't want steak. This resulted in an argument, because she had saved so hard to buy him that stinking steak and he was so unappreciative. I don't think she ever cooked him a steak again for as long as I can remember."
11. The stories in this paragraph come from two of the books about Camp Century written at the time: Walter Wager, *Camp Century: City under the Ice* (Philadelphia: Chilton Books, 1963), and Charles Michael Daugherty, *City under the Ice: The Story of Camp Century* (New York: Macmillan, 1963).
12. This story, like many others about Camp Century, is told in a chapter of Willi Dansgaard's autobiography, *Frozen Annals: Greenland Ice Cap Research* (Odder, Denmark: Narayana Press, 2004), available at https://nbi.ku.dk/english/research/pice/Wiili_Dansgaard_-_Gr_nland_i_istid_og_nutid_-_trykkvalitet.pdf.
13. Two different publications by Walter Wager tell this story similarly: Walter Wager, "Life Inside a Glacier," *Saturday Evening Post*, September 10, 1960, 61; and Walter Wager, *Camp Century: City under the Ice* (Philadelphia: Chilton Books, 1963), 75–76.
14. The Army drafted William Vest, a geologist, and assigned him to Fort Belvoir, in Virginia. From there, he traveled to Greenland several times in the early 1960s. He remembers that at least one woman came to Camp Century when he was there. He believes she was a legislator who flew in, inspected the camp, and flew out the same day. He referred to her as a "bluebird," the name the soldiers gave to visiting dignitaries in an oblique reference to the school buses that transported people around Thule.
15. Wager, *Camp Century*, 75.
16. "Two Scouts Return from Arctic Trip," *New York Times*, April 7, 1961.
17. Henri Bader, "The Problem of Arctic Research," in *Proceedings of the Third*

Annual Arctic Planning Session, November 1960, GRD Research Notes 55 (Bedford, MA: Geophysical Research Directorate, 1961), 109–13. By the time of this meeting, Bader had left the Army's direct employment, and it seems that in this four-page speech, he laid out a lot of his own views of both the benefits and drawbacks of military-supported research.

18. Lists of ham radio call signs from the 1950s and 1960s indicate that call signs for military ham radio stations in Greenland began with KG. KG1CC was the call sign for the Camp Century station. KG1BA was the call sign for the station at Camp Tuto. Thule Air Base was busier. So far, I've found the Thule postcards for KG1AA, KG1BA, KG1CD, KG1BB, KG1CJ, KG1BO, KG1CQ, KG1CB, KG1AR, and KG1DL—mostly on eBay, where they list for a few dollars each but usually include the "make an offer" button. The Thule call signs show up repeatedly in magazines of the 1950s and 1960s. *Shortwave Magazine* and *Popular Electronics* include lists of stations contacted and competitions for the number of countries a call sign reached.
19. The following magazine articles feature Camp Century: Johansen, "U.S. Army Builds a Fantastic City under Ice"; Wager, "Life Inside a Glacier"; "Le Bon Home De Neige," *Spirou*, January 11, 1962; Henri Dimpre, "Camp Century, Cité Polaire," *Pilote*, March 29, 1962; Kent Goering, "Report from Camp Century, Eagle under the Ice," *Boys' Life*, March 1961; and George J. Dufek, "Nuclear Power for the Polar Regions," *National Geographic*, May 1962. These three books are about the camp: Wager, *Camp Century*; Lee David Hamilton, *Century: Secret City of the Snows* (New York: Putnam, 1963); and Daugherty, *City under the Ice.*
20. "Le Bon Home De Neige."
21. Dimpre, "Camp Century, Cité Polaire."
22. *Camp Century—City under the Ice*, color film, U.S. Army Research and Development, Progress Report no. 6, 1961. Langway appears at 29:48.
23. Episode 520 of *The Big Picture*, titled "The U.S. Army and the Boy Scouts," opens with three minutes of the Scouts at Tuto, taking a heavy swing to Camp Century, doing science, and watching the nuclear reactor. *Boys' Life* gave them a color spread (Goering, "Report from Camp Century"). It included a photo of the Scouts ducking beneath a Caution Radiation Area sign as they left the reactor trench. When the boys returned to the States in 1961, the *New York Times* led its April 7 story ("Two Scouts Return from Arctic Trip") this way: "Ninety-seven days of darkness, temperatures 67 degrees below zero and the nearest girls 180 icebound miles away were some of the memories of an Arctic winter brought back yesterday by two 18-year-old Boy Scouts."
24. The first section of *To Tell the Truth*, season 5, episode 31, 1961, features the Greenland story and the Scouts. The episode is available on YouTube and includes accurate descriptions of Camp Century.
25. *TV Guide* (for Sunday, January 15, 1961) described the 24-minute film concisely: "Walter Cronkite visits Camp Century Greenland. 'City under Ice.' 800 miles from the North Pole where early research and early missile warning activities are carried out." In the film, Cronkite rides in

a Weasel with the base commander, shows viewers the base barracks (along with the dog, Mukluk), and documents the reactor startup.

26. Daugherty, *City under the Ice.*
27. A thorough political consideration of Camp Century in the context of the Cold War is provided by Kristian H. Nielsen, Henry Nielsen, and Janet Martin-Nielsen, "City under the Ice: The Closed World of Camp Century in Cold War Culture," *Science as Culture* 23, no. 4 (October 2, 2014): 443–64, https://doi.org/10.1080/09505431.2014.884063.
28. Diesel is a fuel named for a German inventor. In 1892, Rudolf Diesel built an engine that ran without spark plugs and was able to use a variety of different fuels; that technology, albeit modernized, is still going strong more than a century later. Diesel is a ubiquitous fuel that is exceptionally rich in carbon, 87 percent by weight (bested only by coal and peat). It's usually derived from crude oil, but coal, tar sands, and even used French fry oil (used to make biodiesel) are useful feedstocks. Diesel isn't one molecule, it's a mix of many, and each has a different shape and size. There are straight molecules, rings, and rings attached to straight molecules. All have a backbone made of carbon atoms (usually four to twelve) with hydrogen atoms hanging off the edges. There is about seven times more carbon by weight in diesel than there is hydrogen. When diesel burns, much of the energy comes from the breaking of bonds between carbon atoms. That means lots of carbon dioxide emitted for each unit of energy rendered from the fuel, even though most diesel engines are more efficient than those running on gasoline. That carbon is ancient, which explains the fossil fuel designation for all but biodiesel. In Greenland, the Army burned diesel fuel that 100 million years ago was tropical swamp muck and green algae.
29. Lawrence H. Suid, *The Army's Nuclear Power Program: The Evolution of a Support Agency* (New York: Greenwood Publishing Group, 1990).
30. *Design Analysis of a Prepackaged Nuclear Power Plant for an Ice Cap Location* (Schenectady, NY: ALCO Products, January 15, 1959), https://doi.org/10.2172/4049083.
31. James W. Barnett, *Construction of the Army Nuclear Power Plant PM-2A at Camp Century, Greenland: Final Report* (Washington, DC: Dept. of the Army, Office of the Chief of Engineers, Nuclear Power Division, 1961).
32. An extensive review of the geopolitics surrounding Project Iceworm is provided by Erik D. Weiss, "Cold War under the Ice: The Army's Bid for a Long-Range Nuclear Role, 1959–1963," *Journal of Cold War Studies* 3, no. 3 (2001): 31–58. The source of the word "Iceworm" is uncertain, but there's a photo from Site 2 that shows a worm with a top hat and reads, "Welcome, Home of the Iceworm, Detachment of 931st Airborne." It dates to 1957 and was taken by Jerry Greelis (https://campcentury.omeka.net/admin/items/show/202). There are iceworms that live on the surfaces of glaciers in the ablation zones.
33. Project Iceworm is reviewed in two sources with citations to other relevant works, some of which appear to remain classified: Nikolaj Petersen, "The Iceman That Never Came: 'Project Iceworm,' the

Search for a NATO Deterrent, and Denmark, 1960–1962," *Scandinavian Journal of History* 33, no. 1 (March 2008): 75–98, https://doi.org/10.1080/03468750701449554; and William C. Baldwin, *The Engineer Studies Center and Army Analysis: A History of the U.S. Army Engineer Studies Center, 1943–1982* (Corporation of Engineers, 1985).

34. Baldwin, *Engineer Studies Center and Army Analysis.*
35. Gunars Abele, *Subsurface Transportation Methods in Deep Snow,* Technical Report 160 (Hanover, NH: Cold Regions Research and Engineering Laboratory, 1965).
36. USNS *Marine Fiddler* (a large ship specifically designed to transport railroad locomotives) carried the PM-2A to Thule.
37. The look on the face of a technician, pictured in Dufek's "Nuclear Power for the Polar Regions," as he lowers a single fuel rod into the reactor core a story below his feet, demonstrates both the importance and the hazard of this moment.
38. The PM-2A and its kin are well described in this illustrated technical document: William R. Corliss, *Power Reactors in Small Packages,* Understanding the Atom (Oak Ridge, TN: U.S. Atomic Energy Commission, Division of Technical Information, 1964).
39. Clark, *Camp Century,* 51.
40. Dagmar Cole, Drummond's daughter, shared with me his deployment list provided by the Army. It shows that he deployed to Greenland three times: first, from April 15 to July 5, 1960; second, from September 1 to December 16, 1960, and for the last time between April 15 and September 8, 1961. The fall 1960 deployment would have been when the reactor was first fueled, started, shut down, and initially modified.
41. Dagmar Cole recalls that her father told the same stories about Greenland and Camp Century over and over until he died at age 87 in 2018.
42. The history of rabies in the Arctic is reviewed in Torill Mørk and Pål Prestrud, "Arctic Rabies—a Review," *Acta Veterinaria Scandinavica* 45, no. 1 (2004): 1, https://doi.org/10.1186/1751-0147-45-1. Associated Press, "Foxes Pass Rabies to Eskimo Dogs," *Salt Lake Tribune,* September 11, 1959, describes the impacts of the disease as it spread to native Greenlanders' dog teams and suggests that the government will compensate the dogs' owners for their losses.
43. Dave Karten, "New Look in Air Force Medicine," *Airman, Official Journal of the Air Force,* August 1960.
44. Among Danish veterinarians that I contacted, none knew the name of a woman veterinarian active in the early 1960s in Greenland.
45. Cole has spent 30 years trying to identify this engineer with no luck. I fared no better.
46. Clark, *Camp Century.*
47. Occupational limits at the time were 5 rem (= 5,000 mrem) per year, so even if the operators were there all year, they would get only 1,200 mrem, 25 percent of the limit but four times the exposure of the average American. Today, occupational limits in the United States are much

lower, 1,000 mrem per year. The limit for the general public was set at 500 mrem in 1960.

48. See page 50 in Clark, *Camp Century*, for a table listing discharges of hot waste (primary loop cooling water) into the hot waste disposal sump in the ice at Camp Century for 1962. I have not found data for limited operations in 1960, nor the data for 1961 and 1963.
49. Jixin Qiao et al., "High-Resolution Tritium Profile in an Ice Core from Camp Century, Greenland," *Environmental Science & Technology* 55, no. 20 (October 19, 2021): 13638–45, https://doi.org/10.1021/acs.est.1c01975. These researchers measured radioactive tritium in a firn core they collected at Camp Century about 500 feet from the reactor vent. They found elevated levels of tritium between 1959 and 1963, with one spike in 1959 and another in 1962 and 1963. Although the latter spike coincides with the time when the reactor was running, that is also the time when the most intensive atmospheric testing occurred at Novaya Zemlya, also in the Arctic.
50. For example, the U.S. military conducted the Castle series of six tests of American thermonuclear weapons from March to May 1954 on Bikini and Enewetak atolls in the South Pacific, as reported in United States Defense Nuclear Agency, *United States Atmospheric Nuclear Weapons Tests: Nuclear Test Personnel Review*, Castle Series, 1954 (Washington, DC: National Technical Information Service, 1982). The tests lofted so much radioactivity into the atmosphere that it circulated around the planet. It also moved into the Northern Hemisphere, where precipitation and dry fall as dust delivered it to the surface. The fallout was easily detected in Greenland's firn, as reported by W. Ambach and W. Dansgaard, "Fallout and Climate Studies on Firn Cores from Carrefour, Greenland," *Earth and Planetary Science Letters* 8, no. 4 (May 1970): 311–16, https://doi.org/10.1016/0012-821X(70)90117-2.

CHAPTER 5: HITTING BOTTOM

1. Henri Bader, *Scope, Problems, and Potential Value of Deep Core Drilling in Ice Sheets*, Special Report (Hanover, NH: U.S. Army Corps of Engineers, Cold Regions Research and Engineering Laboratory, 1962).
2. Quotation and much of the information about the thermal drill and other drilling at Camp Century comes from an oral history interview with Herbert T. Ueda, conducted by Brian Shoemaker on October 23, 2002, as part of the Ohio State University's Polar Oral History program.
3. Lee David Hamilton, "Century: City under Greenland's Ice," *Science Digest*, May 1962, 61–68.
4. "Thermal Drill," *Science Digest*, June 1961.
5. "Thermal Drill."
6. Much of the technical information about the drill comes from Herbert T. Ueda and Donald E. Garfield, *Drilling through the Greenland Ice Sheet*, Special Report 126 (Hanover, NH: Cold Regions Research and Engineering Laboratory, 1968).

7. Crary was the first person to set foot on both the South and North Poles. Every bit as much a polar explorer as a scientist, he led the seventh expedition ever to reach the South Pole, arriving there in February 1961, just as the Camp Century reactor began running continuously. His obituary tells a story that seems to fit the man: in 1958, while in Antarctica, the icy sea cliff on which he was standing gave way and dropped him into the ocean. Crary spent ninety minutes clinging to a floating chunk of ice before rescue came. Edward Hudson, "Albert Crary, Geophysicist, Dies; An Explorer of Both Polar Regions," *New York Times*, October 31, 1987.
8. Hamilton, "Century: City under Greenland's Ice."
9. At the end of the 1962 summer drilling season (see Bader, *Scope, Problems, and Potential Value of Deep Core Drilling in Ice Sheets*), the Army had planned to move the thermal drill to Antarctica and penetrate more than two miles of ice at Byrd Station, reaching the bottom of the Antarctic Ice Sheet. Without the drill, they scuttled the plan. It eventually happened, seven years later.
10. Pyeong Moon Jang et al., "Peripheral Facial Palsy Caused by Trichloroethylene Vapor Exposure," *Journal of the Korean Society of Emergency Medicine* 19, no. 4 (n.d.): 438–42.
11. The outer diameter of the drill bit on the mechanical drill was 6⅛ inches (0.51 feet), according to Ueda and Garfield. *Drilling through the Greenland Ice Sheet.* The volume of the borehole was therefore 0.255² × 3.1415 × 4,562 feet = 932 cubic feet, or 6,952 gallons.
12. Removal and remediation guidelines for TCE according to the U.S. EPA are given in Hugh H. Russell, John E. Matthews, and Guy W. Sewell, *TCE Removal from Contaminated Soil and Ground Water*, Ground Water Issue (Ada, OK: United States Environmental Protection Agency, Office of Research and Development, Office of Solid Waste and Emergency Response: Superfund Technology Support Center for Ground Water, Robert S. Kerr Environmental Research Laboratory, 1992).
13. The history of atomic bomb testing is well described by "The Years of Atmospheric Testing: 1945–1963," Trinity Atomic Web Site, https://www.abomb1.org/atmosphr/. The United States ceased testing in the atmosphere in 1963. France stopped in 1974, and China ceased atmospheric testing in 1980.
14. Minoru Koide et al., "Characterization of Radioactive Fallout from Pre- and Post-Moratorium Tests to Polar Ice Caps," *Nature* 296, no. 5857 (April 1982): 544–47, https://doi.org/10.1038/296544a0.
15. Minoru Koide et al., "Transuranic Depositional History in South Greenland Firn Layers," *Nature* 269, no. 5624 (September 1977): 137–39, https://doi.org/10.1038/269137a0.
16. "Camp Century, City under Ice, Operational," *Army Research and Development*, January 1961.
17. The filtered material led to at least four peer-reviewed papers, several dealing with the same isotopes we used to date the sub-ice sediment: Paul W. Hodge, Frances W. Wright, and Chester C. Langway, "Stud-

ies of Particles for Extraterrestrial Origin: 3. Analyses of Dust Particles from Polar Ice Deposits," *Journal of Geophysical Research* 69, no. 14 (July 15, 1964): 2919–31, https://doi.org/10.1029/JZ069i014p02919; Chester C. Langway and Ursula B. Marvin, "Some Characteristics of Black Spherules," *Annals of the New York Academy of Sciences* 119, no. 1 (1964): 205–23, https://doi.org/10.1111/j.1749-6632.1965.tb47434.x; R. McCorkell, E. L. Fireman, and C. C. Langway, "Aluminum-26 and Beryllium-10 in Greenland Ice," *Science* 158, no. 3809 (December 29, 1967): 1690–92, https://doi.org/10.1126/science.158.3809.1690; and Edward L Fireman and Chester C Langway, "Search for Aluminum-26 in Dust from the Greenland Ice Sheet," *Geochimica et Cosmochimica Acta* 29, no. 1 (January 1965): 21–27, https://doi.org/10.1016/0016-7037(65)90074-8.

18. "In Memory of Henrik B. Clausen," *News on Geophysics*, Niels Bohr Institute, August 29, 2013.
19. Willi Dansgaard, *Frozen Annals: Greenland Ice Cap Research* (Odder, Denmark: Narayana Press, 2004).
20. An entertaining story of both Patterson's work and lead in the environment is presented in Lucas Reilly, "The Most Important Scientist You've Never Heard Of," *Mental Floss*, May 17, 2017.
21. Claire Patterson, "Age of Meteorites and the Earth," *Geochimica et Cosmochimica Acta* 10, no. 4 (October 1956): 230–37, https://doi.org/10.1016/0016-7037(56)90036-9.
22. All the methods, results, and implications are described in excruciating detail in a 47-page paper by Patterson's team. They collected samples in 1965. The paper was published four years later as M. Murozumi, Tsaihwa J. Chow, and C. Patterson, "Chemical Concentrations of Pollutant Lead Aerosols, Terrestrial Dusts and Sea Salts in Greenland and Antarctic Snow Strata," *Geochimica et Cosmochimica Acta* 33, no. 10 (October 1969): 1247–94, https://doi.org/10.1016/0016-7037(69)90045-3.
23. Murozumi, Chow, and Patterson, "Chemical Concentrations of Pollutant Lead Aerosols."
24. The entire story is told in Joseph R. McConnell et al., "Lead Pollution Recorded in Greenland Ice Indicates European Emissions Tracked Plagues, Wars, and Imperial Expansion during Antiquity," *Proceedings of the National Academy of Sciences* 115, no. 22 (May 29, 2018): 5726–31, https://doi.org/10.1073/pnas.1721818115, as well as in the widespread press coverage of the findings that followed.
25. The email exchange among members of the USMA class of 1955 is available at "Camp Century Greenland," West Point 1955 (website), June 2013, https://www.west-point.org/class/usma1955/D/Hist/Century.htm.
26. Franklin and a co-author tell the official story of the removal in three pages of detail, including six photographs and diagrams, in "Removal of a Nuclear Reactor Core—Camp Century," *Military Engineer* 56 (September 1964): 328–30. The unofficial story is hinted at in banter online between Franklin and Lee McKinney (a West Point classmate). It includes a line suggestive of things happening at Century during the decommissioning that are not in the official record: "A lot of the

story must go untold, for several reasons, and I think Joe [Franklin] will agree." "Camp Century Greenland," West Point 1955.

27. A 1965 Army newsreel shows the reactor removal process: *Relocation of the PM-2A Nuclear Power Plant: MF5 5118* (U.S. Army Corps of Engineers, U.S. Army Engineer Reactors Group, 1965).
28. Joseph P. Franklin, "Removal of the PM-2A Nuclear Power Plant from Camp Century," *Transactions of the American Nuclear Society* 8 (1965): 117–18.
29. A long report (135 pages) details the 1966 testing (and radioactivity) of the PM-2A reactor in Idaho: D. R. Mousseau et al., *PM-2A Reactor Vessel Test Program: Final Report* (Idaho Falls, ID: Idaho Nuclear Corporation National Reactor Testing Station, January 1, 1967).
30. Mousseau et al., *PM-2A Reactor Vessel Test Program.*
31. In Bethesda, Franklin spent time with exquisitely sensitive radiation detectors. First, he would have showered, scrubbed, and gowned to remove any radiation on the outside of his body. Then, he would have sat or lain motionless, perhaps for many minutes, even hours, in rooms built from iron plates six inches thick—thick enough to absorb stray neutrons from cosmic rays—as he was scanned. The Navy salvaged the metal used to build the radiation detection rooms from vessels built before 1945. They used this old iron because new iron (made or recycled after 1945) would contain radioactivity released into the environment from nuclear weapons tests and nuclear reactors.
32. CEF, "Radiation Exposure Evaluation Laboratory Established at National Naval Medical Center," *Naval Research Reviews,* December 1960, 15–18. This article is one of three that are useful to understand the idea of whole body counting. A much broader overview of the science of whole body counting is provided in Frederick W. Lengemann and John H. Woodburn, *Whole Body Counters,* Understanding the Atom (Oak Ridge, TN: U.S. Atomic Energy Commission, Division of Technical Information, 1967). The Bethesda facility, its construction, and the processes it uses are described in a document acquired from the National Archives and Records Administration (NARA II) in College Park, Maryland: E. Richard King, J. S. Burkle, and G. L. Lewis, "An Historical Record of the Radiation Exposure Evaluation Laboratory," Radiation Exposure Evaluation Laboratory (Bethesda, MD: U.S. Naval Hospital, National Naval Medical Center, n.d.).
33. Elmer F. Clark, *Camp Century: Evolution of Concept and History of Design Construction and Performance,* CRREL Technical Report 174 (Hanover, NH: Cold Regions Research and Engineering Laboratory, 1965).
34. Crawford Henderson, "Machines for Snow-Tunnel Maintenance," *Military Engineer* 55, no. 364 (1963): 104–5, describes a variety of mechanized tools used to deal with snow creep closing tunnels as well as the icing of tunnel floors. Henderson's award is mentioned in this article, which includes a head-and-shoulders photograph: "Engineer Cited for Designing Snow Tunnel Trimmer," *Army Research and Development,* March 1963.

35. "Snow Shaver Undergoing Greenland Tunnel Tests," *Army Research and Development*, September 1962.
36. "Armais Arutunoff, Class of 1974," Oklahoma Hall of Fame, Gaylord-Pickens Museum (website), 1974.
37. Armais Arutunoff, Drilling apparatus, U.S. Patent US2654572A, filed 1949 and issued 1953.
38. Ueda and Garfield, *Drilling through the Greenland Ice Sheet.*
39. Pavel Talalay et al., "Antarctic Subglacial Drilling Rig: Part II, Ice and Bedrock Electromechanical Drill (IBED)," *Annals of Glaciology* 62, no. 84 (April 2021): 12–22, https://doi.org/10.1017/aog.2020.38.
40. Lyle B. Hansen and C. C. Langway Jr., "Deep Core Drilling in Ice and Core Analyses at Camp Century, Greenland, 1961–1966," *Antarctic Journal of the United States* 1 (September–October 1966): 207–8.
41. Jørgen Peder Steffensen gave us scans of the logs, which include numerous dated notes from the time the core was stored at CRREL (early 1970s; these logs mostly told where subsamples of the ice were sent, including to "Ohio State, France, Swiss, SR"). Langway must have sent the original logs, along with the ice and frozen sediment, to Copenhagen. The sheets are three-hole-punched and yellowed, with mimeographed columns and headings. They are written in graphite pencil. Red checks, possibly grease pencil and over the original writing, suggest that the sheets were used for inventory. On the last page, at the depth where the sub-ice sediment starts, there is a note in blue ink stating that permacrete was "cut into 10 cm [4 inch] pieces 10/13/72."
42. These core tube numbers are what allowed us to identify the sub-ice samples we analyzed. When these longer sections of core were cut into four-inch pieces in 1972, letters and numbers were added after the core tube designation.
43. Ueda, interview by Brian Shoemaker.
44. Ueda, interview by Brian Shoemaker.
45. Ueda, interview by Brian Shoemaker.
46. The date is reported as July 4 in news stories of the time (including "Army Serves Drink Cooled With Ice 2,000 Years Old," *New York Times*, September 22, 1966). The July 4 date is also reported in Hansen and Langway, "Deep Core Drilling in Ice and Core Analyses," and by Ueda in his oral history interview recorded by Brian Shoemaker in 2002. But Langway's last paper just dates the core completion to "July": Chester C. Langway, *The History of Early Polar Ice Cores*, ERDC/CRREL Technical Report (Hanover, NH: U.S. Army Corps of Engineers, Engineer Research and Development Center, Cold Regions Research and Engineering Laboratory, 2008). Perhaps this date change was born of a desire to align the core's completion date with Independence Day, or the drilling log was filled out incorrectly.
47. "Army Serves Drink Cooled With Ice 2,000 Years Old."
48. I found fifty-four newspapers that ran the photograph, a short story, or both. *Time* covered the Pentagon ice-core unveiling in "Geophysics: History on the Rocks," *Time*, September 30, 1966.

49. James A. Bender, "Lucybelle Bledsoe—1923–1966," *Journal of Glaciology* 6, no. 47 (1967): 755–56, https://doi.org/10.3189/S0022143000020001.
50. Glenn Walker, a Black man and SIPRE employee, cuts ice in a 1953 photo titled "Glenn Walker operates a petrographic saw at the Snow, Ice, and Permafrost Research Establishment," held by the Chicago History Museum and available online. On November 3, 1960, the *Chicago Tribune* published a photograph of Black physicist Earl Shaw. It shows him in the SIPRE cold room removing an ice core from a rack holding over a hundred aluminized core tubes.
51. Lee David Hamilton, *Century: Secret City of the Snows* (New York: Putnam, 1963).
52. Issues with mental health, including suicide, appear to increase with latitude in Greenland: see Karin S. Björkstén, Daniel F. Kripke, and Peter Bjerregaard, "Accentuation of Suicides but Not Homicides with Rising Latitudes of Greenland in the Sunny Months," *BMC Psychiatry* 9, no. 1 (December 2009): 20, https://doi.org/10.1186/1471-244X-9-20.
53. Hamilton, *Century: Secret City of the Snows.*
54. In the summer of 2021, a new book appeared, written by two Danish authors with a long history of thinking about Camp Century: Kristian Hvidtfelt Nielsen and Henry Nielsen, *Camp Century: The Untold Story of America's Secret Arctic Military Base under the Greenland Ice*, English edition (New York: Columbia University Press, 2021). The book included translations of Gregersen's diary, originally written in Danish, which put the Scouts' experience at Camp Century in a vastly different light.
55. Ueda, interview by Brian Shoemaker.

CHAPTER 6: AGING

1. *Camp Century Revisited: A Pictorial View—June 1969* (Hanover, NH: U.S. Army Cold Regions Research and Engineering Laboratory, 1970) is an otherworldly glimpse into the abandoned camp, filled with nearly one hundred images of the camp then (1960) and now (1969). It's striking just how quickly the walls and ceilings have deformed, crushing buildings and bending pipes.
2. Willi Dansgaard, *Frozen Annals: Greenland Ice Cap Research* (Odder, Denmark: Narayana Press, 2004).
3. Dansgaard, *Frozen Annals.*
4. N. S. Gundestrup et al., "Camp Century Survey 1986," *Cold Regions Science and Technology* 14, no. 3 (October 1987): 281–88, https://doi.org/10.1016/0165-232X(87)90020-6; William Colgan et al., "Sixty Years of Ice Form and Flow at Camp Century, Greenland," *Journal of Glaciology* 69, no. 276 (August 2023): 919–29, https://doi.org/10.1017/jog.2022.112.
5. Willi Dansgaard et al., "One Thousand Centuries of Climatic Record from Camp Century on the Greenland Ice Sheet," *Science* 166, no. 3903 (October 17, 1969): 377–81, https://doi.org/10.1126/science.166.3903.377.
6. Henri Bader, *Scope, Problems, and Potential Value of Deep Core Drilling in Ice Sheets*, Special Report (Hanover, NH: U.S. Army Corps of Engineers,

Cold Regions Research and Engineering Laboratory, 1962); Miller, "Instruments and Methods."

7. Samuel Epstein and Robert Sharp, "Oxygen Isotope Studies," *National Academy of Sciences, IGY Bulletin #21 (Transactions, American Geophysical Union)* 40, no. 1 (March 1959): 81–84; W. Dansgaard, "The Abundance of O^{18} in Atmospheric Water and Water Vapour," *Tellus* 5, no. 4 (November 1953): 461–69, https://doi.org/10.1111/j.2153-3490.1953.tb01076.x.
8. Dansgaard raises several issues in the 1969 paper (Dansgaard et al., "One Thousand Centuries of Climatic Record") that prevent him from accurately calculating past temperatures: "(i) the deeper layers of ice in the core originated further inland, where perhaps slightly different climatic conditions existed; (ii) the isotopic composition of seawater, which provides the moisture for the precipitation, changed over time; (iii) the ratio of summer to winter precipitation possibly changed over time; (iv) the predominant wind patterns over Camp Century possibly changed over time; (v) the flow pattern of the ice possibly changed; and (vi) the thickness of the ice sheet changed over time."
9. Paul S. Wilcox et al., "Exceptional Warmth and Climate Instability Occurred in the European Alps during the Last Interglacial Period," *Communications Earth & Environment* 1, no. 1 (December 8, 2020): 57, https://doi.org/10.1038/s43247-020-00063-w.
10. Andrea Dutton, Alexandra Villa, and Peter M. Chutcharavan, "Compilation of Last Interglacial (Marine Isotope Stage 5e) Sea-Level Indicators in the Bahamas, Turks and Caicos, and the East Coast of Florida, USA," *Earth System Science Data* 14, no. 5 (May 19, 2022): 2385–99, https://doi.org/10.5194/essd-14-2385-2022.
11. Victor J. Polyak et al., "A Highly Resolved Record of Relative Sea Level in the Western Mediterranean Sea during the Last Interglacial Period," *Nature Geoscience* 11, no. 11 (November 2018): 860–64, https://doi.org/10.1038/s41561-018-0222-5.
12. C. V. Murray-Wallace et al., "Last Interglacial (MIS 5e) Sea-Level Determined from a Tectonically Stable, Far-Field Location, Eyre Peninsula, Southern Australia," *Australian Journal of Earth Sciences* 63, no. 5 (July 3, 2016): 611–30, https://doi.org/10.1080/08120099.2016.1229693.
13. A. Dutton and K. Lambeck, "Ice Volume and Sea Level during the Last Interglacial," *Science* 337, no. 6091 (July 13, 2012): 216–19, https://doi.org/10.1126/science.1205749.
14. Information in this paragraph is presented in Dansgaard et al., "One Thousand Centuries of Climatic Record." Some of the dates have changed in the past 50 years as dating of past climate events has improved.
15. Dansgaard ends his 1969 paper with two simple but powerful sentences: "It appears that ice-core data provide far greater, and more direct, climatological detail than any hitherto known method. Furthermore, ice cores from some dry-snow zones can provide continuous sedimentary records spanning perhaps several hundred millennia." He was absolutely right.

16. Piero Zennaro et al., "Europe on Fire Three Thousand Years Ago: Arson or Climate? Holocene Fire Activity in the NEEM Core," *Geophysical Research Letters* 42, no. 12 (June 28, 2015): 5023–33, https://doi.org/10.1002/2015GL064259.
17. Joseph R. McConnell et al., "Hemispheric Black Carbon Increase after the 13th-Century Māori Arrival in New Zealand," *Nature* 598, no. 7879 (October 7, 2021): 82–85, https://doi.org/10.1038/s41586-021-03858-9.
18. Susan L. Herron and C. C. Langway, "Derivation of Paleoelevations from Total Air Content of Two Deep Greenland Ice Cores," in *The Physical Basis of Ice Sheet Modeling*, E. D. Waddington and J. S. Walder, eds., IAHS Publication 170 (International Association of Hydrological Sciences, 1987): 283–95.
19. The details of the Vesuvius record in Greenland's ice are in C. Barbante et al., "Greenland Ice Core Evidence of the 79 AD Vesuvius Eruption," *Climate of the Past* 9, no. 3 (June 13, 2013): 1221–32, https://doi.org/10.5194/cp-9-1221-2013.
20. Giuseppe Mastrolorenzo et al., "Herculaneum Victims of Vesuvius in AD 79," *Nature* 410, no. 6830 (April 2001): 769–70, https://doi.org/10.1038/35071167.
21. D. Dahl-Jensen et al., "Past Temperatures Directly from the Greenland Ice Sheet," *Science* 282, no. 5387 (October 9, 1998): 268–71, https://doi.org/10.1126/science.282.5387.268.
22. William K. Klingaman and Nicholas P. Klingaman, *The Year without Summer: 1816 and the Volcano That Darkened the World and Changed History* (New York: St. Martin's Press, 2013) describes in great detail the effects of the Tambora eruption around the world in terms of the cooling and odd weather caused by the eruption.
23. H. B. Clausen and C. U. Hammer, "The Laki and Tambora Eruptions as Revealed in Greenland Ice Cores from 11 Locations," *Annals of Glaciology* 10 (1988): 16–22, https://doi.org/10.3189/S0260305500004092.
24. Quote from C. U. Hammer, H. B. Clausen, and W. Dansgaard, "Greenland Ice Sheet Evidence of Post-Glacial Volcanism and Its Climatic Impact," *Nature* 288, no. 5788 (November 1980): 230–35, https://doi.org/10.1038/288230a0. The paper lists all the major eruptions detected by increased acidity in the Camp Century ice core.
25. Joseph R. McConnell et al., "Extreme Climate after Massive Eruption of Alaska's Okmok Volcano in 43 BCE and Effects on the Late Roman Republic and Ptolemaic Kingdom," *Proceedings of the National Academy of Sciences* 117, no. 27 (July 7, 2020): 15443–49, https://doi.org/10.1073/pnas.2002722117.
26. McConnell et al., "Extreme Climate after Massive Eruption."
27. Robert J. Delmas, Jean-Marc Ascencio, and Michel Legrand, "Polar Ice Evidence That Atmospheric CO_2 20,000 Yr BP was 50% of Present," *Nature* 284, no. 5752 (March 13, 1980): 155–57, https://doi.org/10.1038/284155a0.
28. A. Neftel et al., "Ice Core Sample Measurements Give Atmospheric

CO_2 Content during the Past 40,000 Yr," *Nature* 295, no. 5846 (January 1982): 220–23, https://doi.org/10.1038/295220a0.

29. Natalia Rybczynski et al., "Mid-Pliocene Warm-Period Deposits in the High Arctic Yield Insight into Camel Evolution," *Nature Communications* 4, no. 1 (March 5, 2013): 1550, https://doi.org/10.1038/ncomms2516.
30. We found only three photographs available of the Camp Century sub-ice sediment, all taken at CRREL before 1972 when someone cut the sediment cores into sections several inches long and placed each in a separate glass cookie jar. David Atwood, an Army Air Corps pilot in World War II and then a CRREL photographer, took them. The photograph with Langway and Hansen that leads this chapter was originally in color, and there is a similar one in black and white. The other shows only the core, in several pieces, in a wooden core box. John Fountain et al., "Evidence of the Bedrock beneath the Greenland Ice Sheet near Camp Century, Greenland," *Journal of Glaciology* 27, no. 95 (1981): 193–97, https://doi.org/10.3189/S0022143000011370.
31. U.S. National Science Foundation award number 7000047.
32. The first data about the sub-ice sediment came in an abstract: C. C. Woo, R. F. Commeau, and C. C. Langway Jr., "Scanning-Electron-Microscope Examination of Sand-Grain Particles from an Ice Core from Camp Century, Northwest Greenland," in *Geological Society of America. Abstracts with Programs*, vol. 8 (1976): 176.
33. W. B. Whalley and C. C. Langway, "A Scanning Electron Microscope Examination of Subglacial Quartz Grains from Camp Century Core, Greenland—A Preliminary Study," *Journal of Glaciology* 25, no. 91 (1980): 125–32, https://doi.org/10.3189/S0022143000010340.
34. Whalley and Langway, "A Scanning Electron Microscope Examination of Subglacial Quartz Grains." There are no further indications in the paper of the depth from which the samples given to Whalley came. When I spoke to Whalley, he didn't remember there being any specific sample identifiers with the samples when Langway sent them to him.
35. In an email conversation with Harwood, I learned that he never went to Buffalo, nor did he meet Langway. He remembers that his doctoral advisor, Peter Webb, met with Tony Gow (from CRREL), and that started the project.
36. David M. Harwood, "Do Diatoms beneath the Greenland Ice Sheet Indicate Interglacials Warmer than Present?," *Arctic* 39, no. 4 (January 1, 1986): 304–8, https://doi.org/10.14430/arctic2092.
37. Harwood, "Do Diatoms beneath the Greenland Ice Sheet Indicate Interglacials Warmer than Present?"
38. By training, Fountain was a volcanologist who studied rocks and their chemical composition, but he built his career studying the behavior of toxic substances in the environment.
39. Fountain et al., "Evidence of the Bedrock beneath the Greenland Ice Sheet."
40. Repeated efforts to contact Langway in 2021 and 2023 were unsuccessful.
41. There are newspaper photographs (dated September 6, 1961) from the

Chicago Sun Times of the ice cores from Little America and Site 2 on racks in Willamette, just before those cores moved to CRREL. At the top of that rack are cores on a shelf labeled "Project Chariot," a name that refers to an early 1960s military endeavor to blast a new deepwater harbor at Cape Thompson, in northern Alaska, using nuclear explosions. SIPRE collected these cores of permafrost using chilled diesel fuel as a drilling fluid to keep the permafrost from melting: G. Robert Lange, *An Investigation of Core Drilling in Perennially Frozen Gravels and Rock*, Technical Report 245 (Hanover, NH: Cold Regions Research and Engineering Laboratory,1973). The coring provided data on the temperature and physical characteristics of the permafrost needed for planning the nuclear explosions that never occurred. However, more than 6,000 gallons of diesel fuel remained behind, contaminating the soil and rock. In 2014, the U.S. government removed 786 tons of diesel-contaminated soil. The total cleanup cost was over $3 million. U.S. Department of Energy, Legacy Management, "Fact Sheet, Chariot, Alaska Site," n.d.

42. Marcel de Quervain and Hans Rothlisberger, "Obituary—Henri Bader (1907–1998)," *Ice: News Bulletin of the International Glaciological Society*, no. 120 (1999).
43. Chester C. Langway Jr, "Ice Core Storage Facility," *Antarctic Journal of the United States* 9, no. 6 (1974): 322–25.
44. The information and quotation in this paragraph come from the author's phone conversation with Bill Vest on September 18, 2023.
45. U.S. National Science Foundation award number 7508512, "Operation of the Central Ice Core Storage Facility and Information Exchange," began on January 1, 1975, and ended August 31, 1991, after providing a total of $1,042,523. An abstract is available from the NSF's Award Search web page.
46. Darren Dopp, "Geology Professor's Work Remains Frozen in Time in His Lab," *Los Angeles Times*, December 4, 1988.
47. Two films show the Buffalo freezers. The on-campus freezer is shown in *In Search of* . . . with Leonard Nimoy, season 2, episode 23, "The Coming Ice Age," 1978, in which Langway seems to subscribe to an idea (now disproven) that the climate was then cooling and that the Earth was headed into another ice age. He appears at 10:32 and talks about rapid cooling of Earth 89,000 years ago, with the campus freezer entrance door and temperature gauges in the background. The inside of the commercial freezer in Buffalo is shown in another film, in which the narrator is dressed in clean-room clothing, presumably to protect the ice cores. The images of the freezer start at 47:46 and show tubes of the Dye-3 ice core on the racks behind the narrator. *After The Warming* (Owings Mills, MD: Maryland Public Television, 1990), available at American Archive of Public Broadcasting.
48. Chiang wrote glowingly of Langway and his work in a letter to the editor of the *Washington Post*, published March 26, 2021, and titled "Scientists at the Core of Climate Research," after the publication of Andrew J. Christ et al., "A Multimillion-Year-Old Record of Greenland Vege-

tation and Glacial History Preserved in Sediment beneath 1.4 Km of Ice at Camp Century," *Proceedings of the National Academy of Sciences* 118, no. 13 (March 30, 2021): e2021442118, https://doi.org/10.1073/pnas.2021442118.

49. Dopp, "Geology Professor's Work Remains Frozen in Time in His Lab."
50. Chester Langway and Erick Chiang, "Central Ice Core Storage Facility and Information Exchange," *Antarctic Journal of the United States* 12, no. 4 (1977): 154–56.
51. Much of the information in the following paragraphs comes from the author's phone conversation with Thompson in June 2023.
52. Zwally and I spoke by phone in July 2022. He had worked at the National Science Foundation in the early 1970s and dealt with Langway in that capacity. In our conversation, Zwally described repeated contentious interactions with Langway, who "did not like to be managed." Zwally told me that Langway's claim to fame was "long term relationships with research groups in Europe" and that "Langway wasn't involved as a science peer but rather as a connection to the ice cores." Specifically, Zwally told me that while the ice cores were at CRREL, "Langway was very controlling of who could get those cores. He had a free hand at that and there were problems with that." Zwally reported that National Science Foundation discussions about a national ice-core facility began in 1972 in order to make ice more widely available for scientific research across the entire research community.
53. A 2005 interview with Thompson in *Rolling Stone* describes Thompson well: "Growing up on a small farm in West Virginia, Thompson studied geology so he could work in the coal industry. But he got sidetracked in grad school, when he examined the first ice core ever extracted by American scientists [Camp Century]. 'You could have knocked me over with a feather the day I discovered, firsthand, that glaciers contain a frozen history of the Earth.' " "The Ice Hunter," in "Warriors and Heroes against Global Warming: Twenty-Five Leaders Who Are Fighting to Stave Off the Planetwide Catastrophe," *Rolling Stone,* November 17, 2005, 92.
54. Lonnie G. Thompson, *Analysis of the Concentration of Microparticles in an Ice Core from Byrd Station, Antarctica,* Institute of Polar Studies Report No. 46, September 1973, 44.
55. Lonnie Thompson told me this story during our phone conversation.
56. He started with a very small grant for fieldwork only from the National Science Foundation (according to Zwally) in the summer of 1974 and secured more funding from the foundation in 1975 (NSF-7515513) for a proposal titled "The Paleoclimate of the Quelccaya Ice Cap, Peru, and Its Relationship to Paleoclimate in High Latitudes."
57. L. G. Thompson and W. Dansgaard, "Oxygen Isotope and Microparticle Studies of Snow Samples from Quelccaya Ice Cap, Peru," *Antarctic Journal of the United States* 10, no. 24 (1975): 24–26.
58. Dopp, "Geology Professor's Work Remains Frozen in Time in His Lab."
59. When I spoke with both Palais and Zwally, they each emphasized this

aspiration of the National Science Foundation. It's echoed by Richard Alley, who suggests that the extensive analytical work done in the GISP2 science trench was the result of limited U.S. laboratory facilities for ice-core analysis. Richard B. Alley, *The Two-Mile Time Machine: Ice Cores, Abrupt Climate Change, and Our Future*, Princeton Science Library (Princeton: Princeton University Press, 2014).

60. "A Chilling Conversation with UB's Ice Core King," *Buffalo News*, April 12, 1989.
61. Palais's work is reported in papers such as Julie M. Palais et al., "Magmatic and Phreatomagmatic Volcanic Activity at Mt. Takahe, West Antarctica, Based on Tephra Layers in the Byrd Ice Core and Field Observations at Mt. Takahe," *Journal of Volcanology and Geothermal Research* 35, no. 4 (December 1988): 295–317, https://doi.org/10.1016/0377-0273(88)90025-X.
62. The Colorado award, for a proposal titled "National Ice Core Curatorial Facility," started on January 1, 1991. The award was $1,662,822 for an initial five-year duration. The abstract (NSF-9016366, available from NSF's Award Search web page) clearly describes the collaborative, open, and national approach to ice-core storage and dissemination.

    > This award supports the establishment of a National Ice Core Curatorial Facility to be located on the grounds of the Denver Federal Center, operated as an extension of the U.S. Geological Survey's Core Research Center (CRC) by the University of Colorado at Boulder and the U.S. Geological Survey. The Facility will provide 50,600 cubic feet of safeguarded core storage at −35 degree Celsius. Refrigerated examination room, holding, staging, and changing areas will be provided as well as instruments and equipment for routine core examination and processing. Additional facilities will be available at the Core Research Center. The Institute of Arctic and Alpine Research, University of Colorado will be responsible for the Facility, and will provide direction and communication with the scientific community. The University will also provide a background of relevant research and educational activities that will enhance the Facility and provide new scientific opportunities. The USGS will provide the physical facilities and staff, operate, and maintain the Facility. The USGS will also act in a curatorial capacity, providing for the processing, cataloging, and distribution of samples—an area in which CRC personnel have over forty man-years of experience.

63. The proposal evaluation process at the U.S. National Science Foundation is based in peer review. Grant funding is not a one-person decision; it's a process. The review process starts with a program manager who picks a half-dozen or more scientists with the right expertise to read and evaluate each proposal—these are the ad hoc reviewers. Then, the Foundation convenes a panel of experts—in the past, usually in Washington, DC, now mostly online. Panel members read the proposals and the ad hoc reviews, and then they talk, argue, and consider each pro-

posal's strengths and weaknesses. At the end, the panel gives advice to the program manager, who, for the ice-core storage facility, was Palais. Palais told me that panel knew that some researchers had found it hard to get samples from the Buffalo facility.

CHAPTER 7: WHEN GREENLAND WAS GREEN

1. For a critical review of this gender disparity, see Christina L. Hulbe, Weili Wang, and Simon Ommanney, "Women in Glaciology, a Historical Perspective," *Journal of Glaciology* 56, no. 200 (2010): 944–64, https://doi.org/10.3189/002214311796406202.
2. There are several abstracts that present a generalized view of these data, but none include the specific information that Dahl-Jensen presented that day. As of 2023, the data remain unpublished.
3. Reviewed in Paul R. Bierman, "Using In Situ Produced Cosmogenic Isotopes to Estimate Rates of Landscape Evolution: A Review from the Geomorphic Perspective," *Journal of Geophysical Research: Solid Earth* 99, no. B7 (July 10, 1994): 13885–96, https://doi.org/10.1029/94JB00459.
4. Lee B. Corbett, Paul R. Bierman, and Dylan H. Rood, "An Approach for Optimizing In Situ Cosmogenic ^{10}Be Sample Preparation," *Quaternary Geochronology* 33 (June 2016): 24–34, https://doi.org/10.1016/j.quageo.2016.02.001.
5. Paul R. Bierman et al., "Mid-Pleistocene Cosmogenic Minimum-Age Limits for Pre-Wisconsinan Glacial Surfaces in Southwestern Minnesota and Southern Baffin Island: A Multiple Nuclide Approach," *Geomorphology* 27, no. 1 (February 1, 1999): 25–39, https://doi.org/10.1016/S0169-555X(98)00088-9.
6. Paul R. Bierman, Alan R. Gillespie, and Marc W. Caffee, "Cosmogenic Ages for Earthquake Recurrence Intervals and Debris Flow Fan Deposition, Owens Valley, California," *Science* 270, no. 5235 (October 20, 1995): 447–50, https://doi.org/10.1126/science.270.5235.447.
7. K. A. Marsella et al., "Cosmogenic ^{10}Be and ^{26}Al Ages for the Last Glacial Maximum, Eastern Baffin Island, Arctic Canada," *Geological Society of America Bulletin* 112, no. 8 (August 1, 2000): 1296–1312, https://doi.org/10.1130/0016-7606(2000)112<1296:CBAAAF>2.0.CO;2.
8. Alice H. Nelson et al., "Using In Situ Cosmogenic ^{10}Be to Identify the Source of Sediment Leaving Greenland," *Earth Surface Processes and Landforms* 39, no. 8 (June 30, 2014): 1087–1100, https://doi.org/10.1002/esp.3565.
9. Lee B. Corbett et al., "Measuring Multiple Cosmogenic Nuclides in Glacial Cobbles Sheds Light on Greenland Ice Sheet Processes," *Earth and Planetary Science Letters* 554 (January 2021): 116673, https://doi.org/10.1016/j.epsl.2020.116673.
10. Paul R. Bierman et al., "Preservation of a Preglacial Landscape under the Center of the Greenland Ice Sheet," *Science* 344, no. 6182 (April 25, 2014): 402–5, https://doi.org/10.1126/science.1249047.
11. Lee B. Corbett, Paul R. Bierman, and Dylan H. Rood, "Constraining

Multi-Stage Exposure-Burial Scenarios for Boulders Preserved beneath Cold-Based Glacial Ice in Thule, Northwest Greenland," *Earth and Planetary Science Letters* 440 (April 2016): 147–57, https://doi.org/10.1016/j.epsl.2016.02.004.

12. Herbert T. Ueda and Donald E. Garfield, *Drilling through the Greenland Ice Sheet*, Special Report 126 (Hanover, NH: Cold Regions Research and Engineering Laboratory, 1968) describes the fluid used in drilling the Camp Century hole.
13. Andrew J. Christ et al., "The Northwestern Greenland Ice Sheet during the Early Pleistocene Was Similar to Today," *Geophysical Research Letters* 47, no. 1 (January 16, 2020), https://doi.org/10.1029/2019GL085176.
14. On a cloudy May morning, 2021, I was paging through hundreds of scanned news stories online, all with Langway's name—mostly the same, one after the next—until the masthead of our local paper, the *Burlington Free Press*, caught my eye. On page 13, there was a list of University of Vermont events open to the public. A note at the bottom of the microfilmed page read, "At 4 p.m. Friday [March 1, 1968], in Room 101, Geology Building, Dr. Chester C. Langway Jr. will lecture on 'Greenland Glaciological Activities' and show a color film on polar glaciology. Dr. Langway is Chief, Snow and Ice Branch, U.S. Army Cold Regions Research and Engineering Laboratory, Hanover, NH." This room was where we processed lake sediment cores in the 1990s.
15. Anders J. Noren et al., "Millennial-Scale Storminess Variability in the Northeastern United States during the Holocene Epoch," *Nature* 419, no. 6909 (October 2002): 821–24, https://doi.org/10.1038/nature01132.
16. As far as we know, the only samples removed were fine-grained material used for Whalley and Harwood's electron microscope and diatom work, the pebbles that Fountain analyzed, the one clast Ueda gave to the museum in Japan, and two sections that were missing when Christ inventoried the material in 2021. There is no indication what happened to the two missing sections.
17. NASA provides the moon rock total at "Lunar Rocks and Soils from Apollo Missions," NASA Curation / Lunar (website), https://curator.jsc.nasa.gov/lunar/. Wasserburg's lab, and those working in it, made many analyses of these moon rocks. G. J. Wasserburg et al., "Outline of a Lunar Chronology," *Philosophical Transactions of the Royal Society of London. Series A, Mathematical and Physical Sciences* 285, no. 1327 (March 31, 1977): 7–22, https://doi.org/10.1098/rsta.1977.0039, summarizes some of their data.
18. The total mass of sub-ice sediment is at least 36.2 kg, but we don't know how much was subsampled and used back in the 1970s and 1980s for the early papers. The mass of the frozen sediment is 31.2 kg, and the mass of the ice-rich layer is 5 kg. So, compared with moon rocks, not very much.
19. This analogy is Christ's.
20. Henri Bader, *Scope, Problems, and Potential Value of Deep Core Drilling in Ice Sheets*, Special Report (Hanover, NH: U.S. Army Corps of Engi-

neers, Cold Regions Research and Engineering Laboratory, 1962); Henri Bader, "Trends in Glaciology in Europe," *Geological Society of America Bulletin* 60, no. 9 (1949): 1309, https://doi.org/10.1130/0016-7606(1949)60[1309:TIGIE]2.0.CO;2.

21. John Fountain et al., "Evidence of the Bedrock beneath the Greenland Ice Sheet near Camp Century, Greenland," *Journal of Glaciology* 27, no. 95 (1981): 193–97, https://doi.org/10.3189/S0022143000011370.
22. David M. Harwood, "Do Diatoms beneath the Greenland Ice Sheet Indicate Interglacials Warmer than Present?," *Arctic* 39, no. 4 (January 1, 1986): 304–8, https://doi.org/10.14430/arctic2092.
23. *Science* sent a reporter, Paul Voosen, who was with us for much of the meeting and reported on the new data. His story appeared soon after: Paul Voosen, "Mud in Storied Ice Core Hints at a Thawed Greenland," *Science* 366, no. 6465 (November 2019): 556–57, https://doi.org/10.1126/science.366.6465.556.
24. Carbon-14 has a half-life of 5,730 years. Although instruments used to detect the isotope are extremely sensitive, so few atoms of ^{14}C remain after 8–10 half-lives that they are impossible to detect above background levels. This effectively sets the upper (oldest) limit of the dating method at 40,000–50,000 years.
25. The details and quotations in this and the following paragraphs are based on Josh Brown's recording of that evening.
26. "In Memory of Henrik B. Clausen," *News on Geophysics,* Niels Bohr Institute, August 29, 2013.
27. Jørgen Peder Steffensen, email correspondence with author, summer 2021.
28. In September 2023, the three University of Vermont graduate students working on the Camp Century sub-ice core went to Copenhagen for an ice-core workshop. They visited the Danish ice-core storage facility and photographed the boxes containing the archived portions of the sub-ice sediment. There were two labels on the gray inner box. The smaller was a shipping label with a "State University of New York at Buffalo" return address. Written on the label in red permanent marker was "Camp Century Sub-Ice Material." There was a larger blue piece of cardstock fastened to the box. It read "ICL Box #15" in bold hand-printed letters. In thinner lettering it read, "C.C. Sub-Ice 11 Jars" as well as "Byrd Volcanic Cores." Written in ink on the cardboard box was "& Byrd Volcanic Ash Samples."
29. Langway knew, from his experience with NSF's increasing oversight, that getting samples to European collaborators from the new Colorado repository could be difficult. The foundation now insists that projects in which ice leaves the United States have American collaborators.
30. According to Palais, rumors swirled that the sub-ice sediment was in Denmark.
31. Andrew J. Christ et al., "A Multimillion-Year-Old Record of Greenland Vegetation and Glacial History Preserved in Sediment beneath 1.4 Km of Ice at Camp Century," *Proceedings of the National Academy of Sciences* 118, no. 13 (March 30, 2021), https://doi.org/10.1073/pnas.2021442118.

32. In 1972, personnel at CRREL cut the sub-ice core into 32 segments. Two segments are missing. One was thawed. Another thawed and was refrozen. The Danes had already cut and sent us the two test samples.
33. Andrew J. Christ et al., "Deglaciation of Northwestern Greenland during Marine Isotope Stage 11," *Science* 381, no. 6655 (July 21, 2023): 330–35, https://doi.org/10.1126/science.ade4248.
34. Polychronis C. Tzedakis et al., "Marine Isotope Stage 11c: An Unusual Interglacial," *Quaternary Science Reviews* 284 (May 2022): 107493, https://doi.org/10.1016/j.quascirev.2022.107493.

CHAPTER 8: ICELESS

1. The relevant example is Superstorm Sandy, a hurricane turned nor'easter that devastated New York in 2012. The storm surge flooded some of the train and subway tunnels in the city as it set records for water levels.
2. "Ocean Warming," Global Climate Change: Vital Signs of the Planet (website), NASA, December 2022.
3. "Ocean Warming."
4. Scott A. Kulp and Benjamin H. Strauss, "New Elevation Data Triple Estimates of Global Vulnerability to Sea-Level Rise and Coastal Flooding," *Nature Communications* 10, no. 1 (October 29, 2019): 4844, https://doi.org/10.1038/s41467-019-12808-z.
5. This is not guesswork: radar and seismic surveys of the ice sheet allow precise calculation of the volume of ice sitting on Greenland and thus the volume of water that will pour into the ocean when the ice melts.
6. Benjamin Strauss, "Coastal Nations, Megacities Face 20 Feet of Sea Rise," Climate Central (website), accessed September 4, 2023; A. Dutton et al., "Sea-Level Rise due to Polar Ice-Sheet Mass Loss during Past Warm Periods," *Science* 349, no. 6244 (July 10, 2015): aaa4019, https://doi.org/10.1126/science.aaa4019.
7. Mika Rantanen et al., "The Arctic Has Warmed Nearly Four Times Faster than the Globe since 1979," *Communications Earth & Environment* 3, no. 1 (August 11, 2022): 1–10, https://doi.org/10.1038/s43247-022-00498-3.
8. Clare Eayrs et al., "Rapid Decline in Antarctic Sea Ice in Recent Years Hints at Future Change," *Nature Geoscience* 14, no. 7 (July 2021): 460–64, https://doi.org/10.1038/s41561-021-00768-3.
9. Sapna Sharma et al., "Loss of Ice Cover, Shifting Phenology, and More Extreme Events in Northern Hemisphere Lakes," *Journal of Geophysical Research: Biogeosciences* 126, no. 10 (October 2021): e2021JG006348, https://doi.org/10.1029/2021JG006348, estimates that lakes freeze eleven days later in the fall, and melt seven days earlier in the spring, than they did a century ago.
10. Eske Willerslev et al., "Ancient Biomolecules from Deep Ice Cores Reveal a Forested Southern Greenland," *Science* 317, no. 5834 (July 6, 2007): 111–14, https://doi.org/10.1126/science.1141758; A. M. Schmidt et al., "A Forest-Meadow Palaeo-Ecosystem in Northwestern Greenland Recov-

ered by Ancient DNA of the Camp Century Ice Core," *Japanese Journal of Palynology*, special issue, 58 (2012): 206; Joerg M. Schaefer et al., "Greenland Was Nearly Ice-Free for Extended Periods during the Pleistocene," *Nature* 540, no. 7632 (December 8, 2016): 252–55, https://doi.org/10.1038/nature20146; Alberto V. Reyes et al., "South Greenland Ice-Sheet Collapse during Marine Isotope Stage 11," *Nature* 510, no. 7506 (June 2014): 525–28, https://doi.org/10.1038/nature13456.

11. Jason E. Box et al., "Greenland Ice Sheet Surface Mass Balance Variability (1988–2004) from Calibrated Polar MM5 Output," *Journal of Climate* 19, no. 12 (June 15, 2006): 2783–2800, https://doi.org/10.1175/JCLI3738.1.
12. Henri Bader, *The Greenland Ice Sheet* (Hanover, NH: U.S. Army Cold Regions Research and Engineering Laboratory, Corps of Engineers, 1961).
13. Bramha Dutt Vishwakarma, "Monitoring Droughts from GRACE," *Frontiers in Environmental Science* 8 (2020), doi.org/10.3389/fenvs.2020.584690.
14. NASA provides excellent visualizations to show the mass loss from Greenland over time at "Greenland Ice Loss 2002–2021," GRACE Tellus (website).
15. Matt Conlen, "Visualizing the Quantities of Climate Change: Ice Loss in Greenland and Antarctica," Global Climate Change: Vital Signs of the Planet (website), NASA, March 9, 2020.
16. The data and calculations are presented in The IMBIE Team, "Mass Balance of the Greenland Ice Sheet from 1992 to 2018," *Nature* 579, no. 7798 (March 12, 2020): 233–39, https://doi.org/10.1038/s41586-019-1855-2.
17. In terms of Olympic swimming pools, one such pool holds 2,500 cubic meters. Each ton of water is a cubic meter. So, 3,900 billion tons of ice is 3,900 billion cubic meters. Dividing by 2,500 gives 1.56 billion pools. Earth's population at the start of 2022 is about 7.75 billion people, and 7.75/1.56 = 4.97.
18. Shfaqat Abbas Khan et al., "Spread of Ice Mass Loss into Northwest Greenland Observed by GRACE and GPS: Northwest Greenland Ice Loss," *Geophysical Research Letters* 37, no. 6 (March 2010), https://doi.org/10.1029/2010GL042460.
19. Google Earth clearly shows the effect of ice loss since the 1960s.
20. 660 feet was the elevation of this place on the USGS topographic map, and that's been the name of the place even since.
21. Michon Scott, "Summer 2012 Brought Record-Breaking Melt to Greenland," climate.gov, NOAA, December 5, 2012.
22. Richard B. Alley and Sridhar Anandakrishnan, "Variations in Melt-Layer Frequency in the GISP2 Ice Core: Implications for Holocene Summer Temperatures in Central Greenland," *Annals of Glaciology* 21 (1995): 64–70, https://doi.org/10.3189/S0260305500015615.
23. Elwyn de la Vega et al., "Atmospheric CO_2 during the Mid-Piacenzian Warm Period and the M2 Glaciation," *Scientific Reports* 10, no. 1 (July 9, 2020): 11002, https://doi.org/10.1038/s41598-020-67154-8.

24. "Past Eight Years Confirmed to Be the Eight Warmest on Record," press release no. 12012023, World Meteorological Organization, January 11, 2023.
25. Leo Sands, "This July 4 Was Hot. Earth's Hottest Day on Record, in Fact," *Washington Post*, July 5, 2023.
26. Astoundingly hot temperature data for playground surfaces in Phoenix are reported in Jennifer K. Vanos et al., "Hot Playgrounds and Children's Health: A Multiscale Analysis of Surface Temperatures in Arizona, USA," *Landscape and Urban Planning* 146 (February 2016): 29–42, https://doi.org/10.1016/j.landurbplan.2015.10.007.
27. Henri Bader, "Request for Photographs of Cordilleran Glaciers," *Journal of Glaciology* 12, no. 66 (1973): 526, https://doi.org/10.3189/S0022143000031993.
28. Tina Schoolmeester and Koen Verbist, *The Andean Glacier and Water Atlas: The Impact of Glacier Retreat on Water Resources* (Paris Arendal: United Nations Educational, Scientific and Cultural Organization GRID-Arendal, 2018).
29. Comparison photographs are available at National Park Service, "Glacier Repeat Photos" (web page), Glacier National Park, Montana.
30. Data are in Romain Hugonnet et al., "Accelerated Global Glacier Mass Loss in the Early Twenty-First Century," *Nature* 592, no. 7856 (April 2021): 726–31, https://doi.org/10.1038/s41586-021-03436-z.
31. The story is told on this 2010 web post by Bayly Turner, "Peru Inventor 'Whitewashes' Peaks to Slow Glacier Melt," phys.org. I've not found any follow-up posts or articles about this approach to managing albedo.
32. Kevin L. Bjella, *Thule Air Base Airfield White Painting and Permafrost Investigation. Phases I–IV*, TR-ERDC/CRREL 13-8 (Hanover, NH: Cold Regions Research and Engineering Laboratory, 2013).
33. Dennis Höning et al., "Multistability and Transient Response of the Greenland Ice Sheet to Anthropogenic CO_2 Emissions," *Geophysical Research Letters* 50, no. 6 (March 28, 2023): e2022GL101827, https://doi.org/10.1029/2022GL101827.
34. Isla H. Myers-Smith et al., "Complexity Revealed in the Greening of the Arctic," *Nature Climate Change* 10, no. 2 (February 2020): 106–17, https://doi.org/10.1038/s41558-019-0688-1.
35. S. A. Pedron et al., "More Snow Accelerates Legacy Carbon Emissions from Arctic Permafrost," *AGU Advances* 4, no. 4 (August 2023): e2023AV000942, https://doi.org/10.1029/2023AV000942.
36. Jørgen Hollesen et al., "Predicting the Loss of Organic Archaeological Deposits at a Regional Scale in Greenland," *Scientific Reports* 9, no. 1 (July 11, 2019): 9097, https://doi.org/10.1038/s41598-019-45200-4.
37. *Secrets of the Ice: The Archaeology of Glaciers and Ice Patches* (Glacier Archaeology Program, 2016) is a short video that explains what glacial archeologists do, and some of their finds. It's available on YouTube.
38. Brice Noël et al., "Rapid Ablation Zone Expansion Amplifies North Greenland Mass Loss," *Science Advances* 5, no. 9 (September 6, 2019): eaaw0123, https://doi.org/10.1126/sciadv.aaw0123.

39. William Colgan et al., "The Abandoned Ice Sheet Base at Camp Century, Greenland, in a Warming Climate: Reconsidering Camp Century," *Geophysical Research Letters* 43, no. 15 (August 16, 2016): 8091–96, https://doi.org/10.1002/2016GL069688. See the supplemental materials for a full accounting of the waste.
40. K. Philberth, "The Disposal of Radioactive Waste in Ice Sheets," *Journal of Glaciology* 19, no. 81 (1977): 607–17, https://doi.org/10.3189/S0022143000215517.
41. Colgan et al., "The Abandoned Ice Sheet Base at Camp Century."
42. Jixin Qiao et al., "High-Resolution Tritium Profile in an Ice Core from Camp Century, Greenland," *Environmental Science & Technology* 55, no. 20 (October 19, 2021): 13638–45, https://doi.org/10.1021/acs.est.1c01975.
43. Hou Xiaolin et al., *Radioactivity in an Ice Core from Camp Century, Greenland* (Radioecology Section, Hevesy Laboratory, Center for Nuclear Technologies, Technical University of Denmark, 2018).
44. "Camp Century Greenland," West Point 1955 (website), June 2013, https://www.west-point.org/class/usma1955/D/Hist/Century.htm.
45. Franklin sent an email report to his classmates in 2016, titled "26 APR 2016 Personal SITREP"—military shorthand for situation report. "I believe it proper for me to report to you that I have been diagnosed with renal cancer in my lungs and bones. . . . I believe all of this was probably started by the radiation I took when (some will recall) I was the CO for the removal of that nuclear power plant from Camp Century on the Greenland Ice Cap, back in the early 1960s. During initial investigations, I personally measured a general radiation field of 2,000 rads, which I did by myself; prohibited all troops from entering the contaminated areas. This took less than two minutes, so I have lived to fight another day. . . . My oncologist tells me there is no cure for cancer; these treatments can reduce some of it and arrest the spread. When these effects first came upon me some three years ago, I lost twenty pounds from my frame, which has always been the same size and shape as when I was a cadet. (Go Army Football!!!)"
46. In that time, Franklin had been a professor at West Point, led troops in Vietnam, and served as the assistant to the chairman of the Joint Chiefs of Staff in the Pentagon. He commanded the cadets at West Point when the first woman matriculated. Franklin found time to author a book, *Building Leaders the West Point Way: Ten Principles from the Nation's Most Powerful Leadership Lab*. In it, he expounds on his experiences in Greenland as a commanding officer.
47. Later analysis suggests it is likely to take more warming, and thus more time, for Camp Century to emerge: Baptiste Vandecrux et al., "Firn Evolution at Camp Century, Greenland: 1966–2100," *Frontiers in Earth Science* 9 (2021), doi.org/10.3389/feart.2021.578978. Geoscientists in the future will be able to determine which paper, if either, was correct.
48. Malcolm Mellor, *Undersnow Structures: N-34 Radar Station, Greenland* (Hanover, NH: U.S. Army Materiel Command, Cold Regions Research and Engineering Laboratory, 1964).

49. The primary cooling water was chemically "demineralized" and then stored in the hot waste tank, which trapped about 75 percent of the radioactivity, before it reached the hot waste disposal sump. Army reports, filed yearly, detail the day-by-day dumping of liquid hot waste and its radioactivity. In 1962, the engineers dumped 47,048 gallons of radioactive water into the ice sheet. This water carried 11.3 millicuries of radiation—about a quarter of what the agreement between the Americans and the Danes allowed. Today, most of that radiation is gone. The short-lived isotopes have long ago decayed away. Long-lived nuclides would remain, but we don't know their original concentration and so can't calculate how much radioactivity might remain.
50. Yucheng Lin et al., "A Reconciled Solution of Meltwater Pulse 1A Sources Using Sea-Level Fingerprinting," *Nature Communications* 12, no. 1 (April 1, 2021): 2015, https://doi.org/10.1038/s41467-021-21990-y.
51. See Figure 2 in Peter U. Clark et al., "Consequences of Twenty-First-Century Policy for Multi-Millennial Climate and Sea-Level Change," *Nature Climate Change* 6, no. 4 (April 2016): 360–69, https://doi.org/10.1038/nclimate2923.
52. G. Erkens et al., "Sinking Coastal Cities," *Proceedings of the International Association of Hydrological Sciences* 372 (November 12, 2015): 189–98, https://doi.org/10.5194/piahs-372-189-2015.
53. Roopinder Tara, "Venice's Tide Barrier Has Already Cost 6 Billion Euros—Will It Work?," Engineering.com, April 14, 2023.
54. "An Engineering Marvel Just Saved Venice from a Flood. What about when Seas Rise?," *Washington Post*, November 26, 2022.
55. Jim Morrison, "After Decades of Waterfront Living, Climate Change Is Forcing Communities to Plan Their Retreat from the Coasts," *Washington Post Magazine*, April 13, 2020.
56. Lisa Rein, "Meet the House Science Chairman Who's Trying to Put Global Warming Research on Ice," *Washington Post*, December 1, 2021; John Abraham, "Lamar Smith, Climate Scientist Witch Hunter," *Guardian*, November 11, 2015.
57. Funding for our research is provided by the U.S. National Science Foundation: "Collaborative Research: A Fossil Ecosystem under the Ice: Deciphering the Glacial and Vegetation History of Northwestern Greenland Using Long-Lost Camp Century Basal Sediment" (award number EAR-OPP-2114629).
58. "The Ice Hunter," in "Warriors and Heroes against Global Warming: Twenty-Five Leaders Who Are Fighting to Stave Off the Planetwide Catastrophe," *Rolling Stone*, November 17, 2005, 92.
59. Benjamin D. DeJong et al., "Pleistocene Relative Sea Levels in the Chesapeake Bay Region and Their Implications for the Next Century," *GSA Today*, August 1, 2015, 4–10, https://doi.org/10.1130/GSATG223A.1.
60. Ben Tracy, "Washington, D.C.'s Tidal Basin Faces Uncertain Future with Rising Waters," CBS News, April 22, 2021.
61. Warren Cornwall, "Washington, D.C., Is Sinking . . . Slowly," *Science*, July 30, 2015, https://doi.org/10.1126/science.aac8938; Bilal M. Ayyub, Haralamb G. Braileanu, and Naeem Qureshi, "Prediction and Impact

of Sea Level Rise on Properties and Infrastructure of Washington, DC," *Risk Analysis* 32, no. 11 (November 2012): 1901–18, https://doi.org/10.1111/j.1539-6924.2011.01710.x.

62. Much of the information in this paragraph comes from Karen A. McKinnon and Isla R. Simpson, "How Unexpected Was the 2021 Pacific Northwest Heatwave?," *Geophysical Research Letters* 49, no. 18 (September 28, 2022): e2022GL100380, https://doi.org/10.1029/2022GL100380.
63. Kirsten Schwarz et al., "Trees Grow on Money: Urban Tree Canopy Cover and Environmental Justice," ed. Steven Arthur Loiselle, *PLOS ONE* 10, no. 4 (April 1, 2015): e0122051, https://doi.org/10.1371/journal.pone.0122051; Jonas Schwaab et al., "The Role of Urban Trees in Reducing Land Surface Temperatures in European Cities," *Nature Communications* 12, no. 1 (November 23, 2021): 6763, https://doi.org/10.1038/s41467-021-26768-w.
64. Kendra Pierre-Louis, "Prisons Aren't Remotely Ready for Extreme Weather," Bloomberg.com, July 8, 2023.
65. "Nearly 90% of U.S. Households Used Air Conditioning in 2020," Today in Energy (website), U.S. Energy Information Administration, May 31, 2022.
66. "The Great Texas Freeze: February 11–20, 2021," National Centers for Environmental Information (NCEI) (web page), February 23, 2023.
67. James P. Kossin et al., "Global Increase in Major Tropical Cyclone Exceedance Probability over the Past Four Decades," *Proceedings of the National Academy of Sciences* 117, no. 22 (June 2, 2020): 11975–80, https://doi.org/10.1073/pnas.1920849117.
68. A recent review of rates of change in the ocean due to climate warming is provided in Carlos Garcia-Soto et al., "An Overview of Ocean Climate Change Indicators: Sea Surface Temperature, Ocean Heat Content, Ocean pH, Dissolved Oxygen Concentration, Arctic Sea Ice Extent, Thickness and Volume, Sea Level and Strength of the AMOC (Atlantic Meridional Overturning Circulation)," *Frontiers in Marine Science* 8 (2021), https://doi.org/10.3389/fmars.2021.642372.
69. A frequently updated web page provides extensive information about hurricanes and the future: Tom Knutson, "Global Warming and Hurricanes: An Overview of Current Research Results," Geophysical Fluid Dynamics Laboratory, NOAA, November 17, 2023.
70. Kevin E. Trenberth et al., "Hurricane Harvey Links to Ocean Heat Content and Climate Change Adaptation," *Earth's Future* 6, no. 5 (May 2018): 730–44, https://doi.org/10.1029/2018EF000825.
71. A forecaster at the National Hurricane Center in Miami described feeling a mixture of fear, curiosity, and excitement as Hurricane Andrew struck his hardened building. Looking outside and noticing the destruction of every palm tree lining U.S. Route 1, he knew that this was a Category 5 storm. When he left work that evening, he saw a colleague's car flipped over a four-foot-high concrete wall bordering the parking lot. That's how much force the storm winds carried. This story comes from a detailed history of Hurricane Andrew's impact on Florida

as well as interviews with forecasters preserved at "Hurricane Andrew's 30th Anniversary," News around NOAA (website), National Weather Service.

72. Marla Schwartz and Megan Linkin, *Hurricane Andrew: The 20 Miles That Saved Miami* (Swiss Reinsurance Company, 2017).
73. This address comes from his customs and immigration card dated July 30, 1960, when he returned from Thule to New Jersey on MATS flight 476/30: 3280 SW 17th Ave, Miami, Florida.
74. Higher estimates come from Rebecca Lindsey, "Climate Change: Global Sea Level," climate.gov, April 19, 2022. Lower estimates come from section B.5.3 in Valérie Masson-Delmotte et al., eds., "Summary for Policymakers," in *Climate Change 2021: The Physical Science Basis. Contribution of Working Group I to the Sixth Assessment Report of the Intergovernmental Panel on Climate Change* (Cambridge: Cambridge University Press, 2021), 3–32.
75. This information, and the information that follows about the Baders and their apartment during the storm, comes from Bader's obituary: Marcel de Quervain and Hans Rothlisberger, "Obituary—Henri Bader (1907–1998)," *Ice: News Bulletin of the International Glaciological Society*, no. 120 (1999).
76. de Quervain and Rothlisberger, "Obituary—Henri Bader."
77. Van Orsdel Funeral Home, "Obituary, Adele Bader," *Miami Herald*, September 5, 2001, 22.
78. Lisa Palladino, "Wallace S. Broecker '53, GSAS '58, 'Grandfather of Climate Science,'" *Columbia College Today*, July 2019.
79. David Farrier, "Our Greatest Libraries Are Melting Away," Opinion, *Washington Post*, April 7, 2001.
80. Janet Martin-Nielsen, "'The Deepest and Most Rewarding Hole Ever Drilled': Ice Cores and the Cold War in Greenland," *Annals of Science* 70, no. 1 (January 2013): 47–70, https://doi.org/10.1080/00033790.2012.721123. Martin-Nielsen illustrates well the transition between what she terms the "military dominated glaciological research" in Greenland and the use of Greenland to pursue questions related to climate.
81. Emma Donoghue, *Astray* (New York: Little, Brown, 2012).
82. Clark et al., "Consequences of Twenty-First-Century Policy."
83. Jessica E. Tierney et al., "Past Climates Inform Our Future," *Science* 370, no. 6517 (November 6, 2020): eaay3701, https://doi.org/10.1126/science.aay3701.
84. An excellent, accessible, and thorough consideration of how long the impact of humans on Earth's climate will last is presented by David Archer, *The Long Thaw: How Humans Are Changing the Next 100,000 Years of Earth's Climate* (Princeton: Princeton University Press, 2016).
85. Henri Bader and Gerald Wasserburg, "Presentation of the Arthur L. Day Medal to Gerald J. Wasserburg," *Geological Society of America Bulletin* 83 (1972): xvii–xxii.
86. Franz Wohlgezogen et al., "The Wicked Problem of Climate Change and Interdisciplinary Research: Tracking Management Scholarship's

Contribution," *Journal of Management & Organization* 26, no. 6 (November 2020): 1048–72, https://doi.org/10.1017/jmo.2020.14.

87. The worry is that Greenland's ice could reach a tipping point where ice loss is not reversible as feedback loops kick in. Alexander Robinson, Reinhard Calov, and Andrey Ganopolski, "Multistability and Critical Thresholds of the Greenland Ice Sheet," *Nature Climate Change* 2, no. 6 (June 2012): 429–32, https://doi.org/10.1038/nclimate1449, suggests that the threshold for losing almost all of Greenland's ice is about 3°F of average global warming. Substantial amounts of ice loss will occur with lower temperature rises.
88. Gabriel Popkin, "Farmers Are Being Paid Millions to Trap Carbon in Their Soils. Will It Actually Help the Planet?," *Science*, July 27, 2023, https://doi.org/10.1126/science.adj9509.

INDEX

Page numbers in *italics* refer to photos and illustrations.
Page numbers after 233 refer to notes.